Group Theory and the Coulomb Problem

GROUP THEORY AND
THE COULOMB PROBLEM

M. J. ENGLEFIELD

Monash University
Victoria, Australia

WILEY-INTERSCIENCE, a Division of John Wiley & Sons, Inc.
New York • London • Sydney • Toronto

LIBRARY
University of Texas
At San Antonio

Preface

For many years group theory has been recognized as an important tool for the theoretical physicist. The three-dimensional rotation group has widespread applications, and most students of quantum mechanics encounter this example, if only in the guise of the quantum theory of angular momentum. Pursuing the subject beyond this, one finds on the one hand books devoted to abstract group theory, especially representation theory, and on the other hand books devoted to specialized applications. There is, in fact, an ample literature available to those specializing in general group-theoretical methods or in special fields in which particular groups are extensively applied. This book is intended for more general readers, who have taken a course in quantum mechanics and would like further acquaintance with group theory beyond their knowledge of angular momentum, without having to venture into unfamiliar and advanced physical fields.

The stimulus for such a text has been the recent interest in the hydrogen atom as an example of symmetry methods. The basic papers on this appeared more than thirty years ago, but the last ten years have seen an upsurge of activity, measured by the number of relevant articles published. Interesting new results have appeared, and the Coulomb problem now seems as good an example of group-theoretical methods as the more complicated (and less understood) physical situations which have already induced monographs. Now the nonrelativistic hydrogen atom surely appears in every first course and every elementary text devoted to quantum mechanics. The application therefore seems ideal for giving general readers an insight into the relation of group theory to quantum mechanics.

Also, there does not seem to be a monograph on this topic. The book should therefore be of value as a reference and as a unified account. I have tried to give proper credit to the originators of the ideas used, but the bibliography is not intended to be complete. My presentation of the noninvariance group follows closely the work of A. O. Barut, but I have explicitly used the relation between the wave functions of the hydrogen atom and those of the two-dimensional harmonic oscillator.

Each chapter begins with introductory paragraphs summarizing its content and giving some idea of its dependence on previous chapters. I hope that many sections can be used independently; for instance, Sections 3.1, 6.1, and

6.2 on harmonic oscillators could be taken consecutively, and Chapters 5 and 6 do not depend on Chapter 4. It is also possible to avoid any explicit group theory and regard the book as an account of algebraic methods based on angular momentum theory. For this approach readers should omit Chapter 1, and Sections 2.4 and 3.7.

M. J. ENGLEFIELD

Melbourne, Australia
August 1971

Contents

1. Lie Algebras and Groups

The purpose of this chapter is to assemble the algebraic theory needed for the Coulomb problem and to illustrate this theory with a variety of examples, many of which are relevant to later chapters. Readers are assumed to be familiar with vectors, matrices, determinants, and those ideas on linear spaces which are used in quantum mechanics: linear dependence, eigenvalues, bases, inner products, and the complex conjugate of an operator. The only special functions used are the $_2F_1$ hypergeometric function and the Gegenbauer and Legendre polynomials.

Underlying an algebraic calculation in quantum mechanics there is usually the Lie algebra of a group rather than the group itself (L66). In view of this I begin with Lie algebras and their representations and then proceed to obtain the groups as exponential functions on the representations. This avoids the differential properties of Lie groups, and I hope it gives a satisfactory illustration of the particular algebras and groups appearing in this book. The different groups corresponding to one Lie algebra are distinguished by the eigenvalues of operators of the algebra.

The list of section headings in the Contents indicates what is covered. Although this chapter is primarily a summary, I include some detail on topics that seem to be treated insufficiently in available texts. Particular examples are the algebra of the pseudorotation group $SO_{2,1}$ and the different kinds of product representation appearing in Section 1.16. On the other hand, I quote without derivation the properties of the hyperspherical harmonics, the four-dimensional generalization of the spherical harmonics. Familiar methods for the spherical harmonics can be extended to obtain these results (K62), and such details would not illuminate the algebraic ideas that are the essential point of the chapter. Actually the later work in the book allows these results to be inferred from solutions of the Coulomb problem.

Finally, this chapter is not an exposition of algebraic theory. It does not even purport to give all the concepts appearing in the literature on the Coulomb problem, but only those used in this book, presented in a way designed to avoid as much theory as possible.

1.1. LIE ALGEBRAS

Suppose $\mathscr{L}$ is a vector space, with elements A, B, etc., with a product $[A, B]$ defined having the following properties:

$$[A, B] \text{ belongs to } \mathscr{L} \text{ for all } A \text{ and } B \tag{1.1}$$

$$[A, B] \text{ is linear in } A \text{ and in } B \tag{1.2}$$

$$[A, A] = 0 \text{ (implying } [A, B] = -[B, A]) \tag{1.3}$$

$$[A, [B, C]] + [B, [C, A]] + [C, [A, B]] = 0 \tag{1.4}$$

Then $\mathscr{L}$ is a Lie algebra. A familiar example is when the elements are ordinary vectors, and $[\mathbf{A}, \mathbf{B}] = \mathbf{A} \times \mathbf{B}$. This is a real Lie algebra, since the vector space is over the field of real numbers. A complex Lie algebra is linear over the field of complex numbers. $[A, B]$ is called the Lie product. The rank of a Lie algebra is the maximum number of independent elements whose mutual Lie products are all zero. Vector algebra has rank 1, since no linearly independent vectors have zero cross product.

The conditions for an algebra to be simple or semisimple are given in Appendix B. They are satisfied for all algebras used in this book. For a simple or semisimple complex Lie algebra of rank l, a basis $H_1, H_2, \ldots,$ H_l; $E_1, E_2, \ldots, E_{n-l}$ can be chosen such that

$$[H_i, E_j] = \alpha_{ij} E_j, [H_i, H_j] = 0 \tag{1.6}$$

where α_{ij} are complex numbers. The square bracket notation for the Lie product is suitable for applications to quantum mechanics, where A and B are operators and AB is the ordinary product. The Lie product is the commutator $AB - BA$, for which the conditions (1.2) through (1.4) are obvious. Any associative algebra of operators thus becomes a Lie algebra. Using the ordinary product gives operators that are not in the Lie algebra. Any such operators which commute with every operator of the Lie algebra are called Casimir operators of the algebra. One example $\sum_{ij=1}^{n} g_{ij} E_i E_j$ always has the form where the numbers g_{ij} are defined in Appendix B.

1.2. THE LIE ALGEBRAS o_3 AND o_4

The real Lie algebra o_n is defined by the basis consisting of the $\frac{1}{2}n(n-1)$ operators

$$D_{\alpha\beta} = -x_\alpha \frac{\partial}{\partial x_\beta} + x_\beta \frac{\partial}{\partial x_\alpha} \quad (\alpha < \beta = 1, 2, \ldots, n) \tag{1.7}$$

with domain chosen so that commutators exist and the partial derivative operators commute. The nonzero Lie products are given by the commutation relations ($D_{\beta\alpha} = -D_{\alpha\beta}$):

$$[D_{\alpha\beta}, D_{\beta\gamma}] = D_{\gamma\alpha} \qquad (1.8)$$

If there is no common subscript the Lie product is zero. The cases $n = 3$ and 4 will be used in this book. Vector algebra and o_3 are isomorphic through the correspondence $(D_{23}, D_{31}, D_{12}) \leftrightarrow (\mathbf{i}, \mathbf{j}, \mathbf{k})$.

The operators used in quantum mechanics act on complex-valued functions which have a complex inner product. With a suitable domain, the operators (1.7) are skew-Hermitian, and because Hermitian operators are physically significant, for $n = 3$ it is usual to consider $L_x = iD_{23}, L_y = iD_{31}$, and $L_z = iD_{12}$ with the commutation relations

$$[L_x, L_y] = iL_z, [L_y, L_z] = iL_x, [L_z, L_x] = iL_y \qquad (1.9)$$

When the operators are Hermitian, the consequences of (1.9) are well known in quantum mechanics and are summarized in the next chapter.

The expression $L^2 = L_x{}^2 + L_y{}^2 + L_z{}^2$ gives a Casimir operator. To get a basis with the property (1.6), L_z is usually chosen for H_1, and $L_\pm = L_x \pm iL_y$ are chosen for E_1 and $E_2 : [L_z, L_\pm] = \pm L_\pm$.

The use of $L_\pm$ implies passing to the algebra $c*o_3$, the complex extension (or complexification) of o_3, consisting of all linear combinations of D_{12}, D_{23}, and D_{31} using complex coefficients. The commutators $[L_z, L_\pm] = \pm L_\pm$, $[L_+, L_-] = 2L_z$ could be used to define a real Lie algebra with basis $L_z, L_\pm$. However, this is not an o_3 algebra, because it cannot be isomorphic to vector algebra: $\mathbf{a} \times \mathbf{b} = \mathbf{b}$ implies $\mathbf{b} = \mathbf{0}$.

For $n = 4$, put $iD_{14} = M_x, iD_{24} = M_y, iD_{34} = M_z$. The Lie products may then be given by

$$[L_x, L_y] = iL_z, [M_x, M_y] = iL_z, [L_x, M_x] = 0,$$
$$[L_x, M_y] = iM_z, [L_x, M_z] = -iM_y \qquad (1.10)$$

and equations obtained by cyclic permutation of x, y, and z. The Casimir operators are $L^2 + M^2$, $\mathbf{L} \cdot \mathbf{M} = L_x M_x + L_y M_y + L_z M_z$, and $\mathbf{M} \cdot \mathbf{L}$, but substituting from (1.7) shows that $\mathbf{L} \cdot \mathbf{M} = \mathbf{M} \cdot \mathbf{L} = 0$. The rank is 2, since L_z and M_z can be chosen for the commuting operators of the basis. Simpler commutation relations are obtained by choosing $K_z = \frac{1}{2}(L_z + M_z)$ and $N_z = \frac{1}{2}(L_z - M_z)$ for H_1 and H_2, and $K_\pm = \frac{1}{2}(L_x + M_x \pm iL_y \pm iM_y)$, $N_\pm = \frac{1}{2}(L_x - M_x \pm iL_y \mp iM_y)$ for the E_j. Their Lie products are

$$[K_z, K_\pm] = \pm K_\pm, [K_z, N_\pm] = [N_z, K_\pm] = 0, [N_z, N_\pm] = \pm N_\pm \qquad (1.11)$$

$$[K_+, K_-] = 2K_z, [N_+, N_-] = 2N_z,]K_\pm, N_\pm] = [K_\pm, N_\mp] = 0$$

where the first line lists the products $[H_i, E_j]$ of (1.6).

Evidently K_z and K_+ span a subalgebra which is isomorphic to $c*o_3$. Also N_z and N_+ span a $c*o_3$ subalgebra, and $c*o_4$ is the direct sum of these two subalgebras. Similarly, the commutators of $-iK_x = \frac{1}{2}(D_{23} + D_{14})$, etc., show that o_4 is the direct sum of two o_3 algebras. Because $\mathbf{L} \cdot \mathbf{M} = \mathbf{M} \cdot \mathbf{L} = 0$, the Casimir operator $L^2 + M^2$ becomes $4K^2 = 4N^2$.

1.3. THE SPHERICAL HARMONICS

Consider any Lie algebra of operators with a basis $H_1, \ldots, E_{n-l}$ as in (1.6) and Casimir operators C_k. For quantum-mechanical applications the elements of the domain of the operators are called states. Since the operators C_k, H_i commute, they have simultaneous eigenstates $|C'; H'\rangle$. The commutators

$$C_k E_j - E_j C_k = 0, \; H_i E_j - E_j H_i = \alpha_{ij} E_j \tag{1.12}$$

imply that if $E_j|C'; H_1', H_2' \cdots H_l'\rangle$ is not zero, then it is another simultaneous eigenstate belonging to the same eigenvalues C' and to the eigenvalues $H_i' + \alpha_{ij}$ of the H_i ($i = 1, 2, \ldots, l$). Explicitly,

$$H_i(E_j|C'; H'\rangle) = E_j(H_i + \alpha_{ij})|C'; H'\rangle = (H_i' + \alpha_{ij})(E_j|C'; H'\rangle)$$

If $\alpha_{ij} > 0$, E_j is often called a raising operator for (eigenvalues of) H_i and is called a lowering operator if $\alpha_{ij} < 0$. Usually the C_k and H_i will be Hermitian, and their eigenvalues and the α_{ij} will be real. Then if E_j is a raising operator, the complex conjugate of (1.12) shows that $E_j^\dagger$ is a lowering operator. In general the α_{ij} appear in pairs of opposite sign.

When the operators $D_{\alpha\beta}$ of the algebra o_n are expressed in spherical polar coordinates, the radial coordinate does not appear. The Casimir operator $\sum_{i<j=1}^n g_{ij} E_i E_j$ is the angular part of $\sum_{\alpha,\beta=1}^n x_\beta^2(\partial^2/\partial x_\alpha^2)$ and its eigenfunctions are called spherical harmonics. The following properties and explicit expressions are a consequence of (1.7), provided that the domain is such that the operators are skew-Hermitian.

For o_3 the functions are (M58, E60)

$$Y_l^m(\theta, \phi) = (-)^l \left[\frac{(2l+1)}{4\pi} \frac{(l+m)!}{(l-m)!}\right]^{1/2} \frac{e^{im\phi}}{l!2^l \sin^m \theta} \frac{d^{l-m}}{(d\cos\theta)^{l-m}} \sin^{2l}\theta \tag{1.13}$$

belonging to the eigenvalues $l(l+1)$ of L^2 and m of L_z, where l is a nonnegative integer and m is one of the $2l+1$ integers $l, l-1, \ldots, -l$. The functions form an orthonormal basis for the angular functions in the domain considered. The factor $(-)^l$ gives the conventional phase to the function. The dependence of this phase on l is arbitrary here, but the dependence on m is fixed by requiring $L_+ Y_l^m$ to be a positive multiple of Y_l^{m+1}. Thus the explicit cases to $l = 4$ given by Bethe and Salpeter (B57) have to be multiplied by $(-)^m$ to agree with (1.13). These three-dimensional spherical harmonics are an example for o_3 of the eigenstates $|C'; H'\rangle$, with $C = L^2$ and $H = L_z$. When $\theta = \frac{1}{2}\pi$, (1.13) is zero if $l - m$ is odd; if $l - m$ is even then

$$Y_l^m(\tfrac{1}{2}\pi, \phi) = (-)^{\frac{1}{2}(l+m)} \frac{[(2l+1)(l+m)!(l-m)!]^{1/2}}{2^{l+1}(\tfrac{1}{2}l + \tfrac{1}{2}m)!(\tfrac{1}{2}l - \tfrac{1}{2}m)! \sqrt{\pi}} e^{im\phi} \tag{1.14}$$

For o_4 the spherical polar coordinates are defined by $x_1 = R \sin \alpha \sin \theta \cos \phi$, $x_2 = R \sin \alpha \sin \theta \sin \phi$, $x_3 = R \sin \alpha \cos \theta$, $x_4 = R \cos \alpha$. Appendix C gives explicitly the operators $D_{\alpha\beta}$ in these coordinates. Since $R \sin \alpha$ is the length of the projection of the four-dimensional vector onto the first three dimensions, the angles θ, ϕ are polar angles in the first three dimensions. M^2 commutes with L^2 and L_z, so that eigenfunctions of $L^2 + M^2$ can be chosen to be also eigenfunctions of L^2 and L_z, and will then depend on θ and ϕ as in (1.13). The resulting four-dimensional spherical harmonics, or hyperspherical harmonics, are the functions (S56)

$$Y_{nlm}(\alpha, \theta, \phi) = i^{n-1-l} 2^{l+1} l! \left[\frac{n(n-l-1)!}{2\pi(n+l)!} \right]^{1/2} \sin^l \alpha$$

$$\times C_{n-l-1}^{l+1}(\cos \alpha) \; Y_l^m(\theta, \phi) \tag{1.15}$$

where C_{n-l-1}^{l+1} is a Gegenbauer polynomial (H65). The Y_{nlm} is an eigenfunction of $L^2 + M^2$ belonging to the eigenvalue $n^2 - 1$, where n is a positive integer. Again l and m are restricted as in (1.13), and also $l < n$, so that for a given n there are n^2 different Y_{nlm}. The phase factor i^{n-1-l} is required for consistency with the conventions of angular momentum theory. Again the functions form an orthonormal basis, satisfying

$$\int_0^\pi \sin^2 \alpha \, d\alpha \int_0^\pi \sin \theta \, d\theta \int_0^{2\pi} d\phi \, \overline{Y}_{n'l'm'} Y_{nlm} = \delta_{n,n'} \delta_{m,m'} \delta_{l,l'} \tag{1.16}$$

For $n \leq 6$ and $m = 0$ these functions are given explicitly by Shibuya and Wulfman (S65a) in the form $\Psi_{nlm} = i^{l+1-n} \pi 2^{1/2} Y_{nlm}$.

As L^2 is not in the algebra o_4, the Y_{nlm} are eigenfunctions of only one member L_z of the algebra. Thus they are not an example for o_4 of the $|C'; H'\rangle$, which will be eigenfunctions of two commuting members of o_4. Using the coordinates

$$x_1 = R \cos \xi \cos \phi, \; x_2 = R \cos \xi \sin \phi, \; x_3 = R \sin \xi \cos \eta,$$
$$x_4 = R \sin \xi \sin \eta$$

eigenfunctions of K^2, K_z, and N_z are given by (S56):

$$Z_{nfg} = (2\pi)^{-1}(2n)^{1/2}(-)^{F+g} d_{gf}^F(2\xi) \exp\{i\phi(f+g) + i\eta(f-g)\} \tag{1.17}$$

where $F = \frac{1}{2}n - \frac{1}{2}$, f and g are the eigenvalues of K_z and N_z, and

$$d_{gf}^F(2\xi) = [(F + f)!(F + g)!(F - g)!(F - f)!]^{1/2}$$

$$\times \sum_t \frac{(-)^t(\cos \xi)^{2F+g-f-2t}(\sin \xi)^{2t-g+f}}{(F + g - t)!(F - f - t)!t!(t - g + f)!} \quad (1.18)$$

Here t is summed over all integral values such that the argument of each of the factorials is nonnegative. The F is a nonnegative integer or half-integer, and f and g can take the $2F + 1$ values $-F$, $-F + 1$, ..., $F - 1$, F. For $0 \leq F \leq \frac{7}{2}$ ($1 \leq n \leq 8$) explicit algebraic forms of (1.18) are listed by Albert (A69). The operators (1.11) are given in Appendix C.

The Y_{nlm} and the Z_{nfg} are alternative sets of basis functions for the n^2-dimensional space of eigenfunctions of $L^2 + M^2$ belonging to the eigenvalue $n^2 - 1$.

The spherical harmonics (1.13) can also be written in terms of the function (1.18):

$$Y_{lm}(\theta, \phi) = \left[\frac{2l + 1}{4\pi}\right]^{1/2} d_{m0}^l(\theta)e^{im\phi}$$

$$= (-)^m\left[\frac{2l + 1}{4\pi}\right]^{1/2} d_{0m}^l(\theta)e^{im\phi} \quad (1.19)$$

1.4. INVARIANT SUBSPACES

Suppose that $\mathcal{S}$ is a subset of the domain of an operator A such that Af is in $\mathcal{S}$ whenever f is in $\mathcal{S}$. Then $\mathcal{S}$ may be called invariant under the operator A. The idea extends to functions invariant under more than one operator, and in particular under a Lie algebra $\mathcal{L}$ of operators.

The last two sections provide several examples. The space spanned by the six functions $x_i x_j$ ($i \leq j = 1, 2, 3$) is invariant under the operators of o_3. This space is the set of homogeneous functions of degree 2 in three variables. More generally homogeneous functions of degree m in n variables form an invariant subspace for the operators (1.7) defining o_n. The spherical harmonics (1.13) of the same l span a $(2l + 1)$-dimensional space invariant under o_3, and the hyperspherical harmonics Y_{nlm} or Z_{nfg} span an n^2-dimensional space $\mathcal{S}_n$ invariant under o_4.

An invariant space may contain a subspace which is invariant and may be the direct sum of such invariant subspaces. Thus the second-degree functions include $r^2 = x_1^2 + x_2^2 + x_3^2$, an eigenfunction of every o_3 operator belonging to the eigenvalue zero. Hence the set $\lambda(x_1^2 + x_2^2 + x_3^2)$ with λ arbitrary is invariant under o_3. The subspace spanned by $x_1{}^2 - x_2{}^2$, $x_2{}^2 - x_3{}^2$, $x_1 x_2$, $x_2 x_3$, $x_3 x_1$ is also invariant. The space of second-degree functions is thus decomposed into the direct sum of two invariant subspaces, bases for which are also given by $r^2 Y_0{}^0$ and $r^2 Y_2{}^m$. On the other hand, the space spanned by the $(2l + 1)$ functions $Y_l{}^m$ has no subspace invariant under all operators of o_3.

A space invariant under a Lie algebra is invariant under any subalgebra. In particular, a space invariant under o_4 will be invariant under three different o_3 subalgebras: one containing D_{12}, D_{23}, and D_{31}, the others containing $\frac{1}{2}D_{23} \pm \frac{1}{2}D_{14}$, $\frac{1}{2}D_{31} \pm \frac{1}{2}D_{24}$, and $\frac{1}{2}D_{12} \pm \frac{1}{2}D_{34}$. If $\mathscr{S}_n$ is decomposed into invariant subspaces of D_{12}, D_{23}, and D_{31}, then the basis Y_{nlm} is appropriate, each different l giving the basis of a $(2l + 1)$-dimensional subspace and $\mathscr{S}_n$ being the direct sum of these n subspaces. The alternative basis Z_{nfg} shows how $\mathscr{S}_n$ is decomposed into n invariant subspaces of either of the other two o_3 subalgebras.

1.5. MATRIX REPRESENTATIONS

Suppose that corresponding to each element A of an algebra is a square matrix M_A, and the correspondence preserves the Lie product, that is $[A, B]$ corresponds to $M_A M_B - M_B M_A$. Then the matrices form a representation $[M]$ of the Lie algebra. As an example, let the matrix

$$\begin{bmatrix} 0 & -A_z & A_y \\ A_z & 0 & -A_x \\ -A_y & A_x & 0 \end{bmatrix}$$

correspond to $\mathbf{A} = A_x \mathbf{i} + A_y \mathbf{j} + A_z \mathbf{k}$, with the components relative to right-handed, orthogonal axes. Then $M_\mathbf{A} M_\mathbf{B} - M_\mathbf{B} M_\mathbf{A}$ corresponds to $\mathbf{A} \times \mathbf{B}$. This also provides a matrix representation of o_3 through the isomorphism $D_{12} \leftrightarrow \mathbf{k}$, etc.

If T is any regular matrix of the same order, then another representation is obtained by letting $T M_A T^{-1}$ correspond to A. All such representations are equivalent. Representations are also termed orthogonal, symmetric, etc., if all the matrices have the named property. If T can be chosen so that each $T M_A T^{-1}$ has the partitioned form

$$\left[\begin{array}{c|c} K_A & 0 \\ \hline 0 & L_A \end{array} \right]$$

where the arrays K_A, L_A are square with order (size) independent of A, then the K_A form a matrix representation, and so do the L_A. The original representation $[M]$ is then called reducible, and the transformation T is said to reduce the representation into the direct sum of $[K]$ and $[L]$. This is written $[M] = [K] + [L]$. If no such transformation T exists, the representation is irreducible.

For ordinary vectors, the representation given above is irreducible. It is also skew-symmetric. A reducible representation of $\mathbf{R} = X\mathbf{i} + Y\mathbf{j} + Z\mathbf{k}$ is

$$M_{\mathbf{R}} = \frac{i}{4} \begin{bmatrix} -2Z & -X + iY & -X + iY & 0 \\ -2X - 2iY & Z - X & Z + X & 2Y - 2iZ \\ -2X - 2iY & Z + X & Z - X & -2Y + 2iZ \\ 0 & Y + iZ & -Y - iZ & 2X \end{bmatrix}$$

This gives a representation, because $M_{\mathbf{R}} M_{\mathbf{S}} - M_{\mathbf{S}} M_{\mathbf{R}}$ corresponds to $\mathbf{R} \times \mathbf{S}$. The representation is reducible, since

$$TM_{\mathbf{R}}T^{-1} = \frac{-i}{2} \left[\begin{array}{cc:cc} Z & X - iY & 0 & 0 \\ X + iY & -Z & 0 & 0 \\ \hdashline 0 & 0 & X & Y - iZ \\ 0 & 0 & Y + iZ & -X \end{array} \right]$$

$$\text{with } T = \begin{bmatrix} 1 & 0 & 0 & 0 \\ 0 & \frac{1}{2} & \frac{1}{2} & 0 \\ 0 & -\frac{1}{2} & \frac{1}{2} & 0 \\ 0 & 0 & 0 & 1 \end{bmatrix} \quad (1.20)$$

The 2×2 matrices $K_{\mathbf{R}}$ are skew-Hermitian and almost unitary: $K_{\mathbf{R}}^{\dagger} = -K_{\mathbf{R}} = \operatorname{adj} K_{\mathbf{R}}$. The $[K]$ and $[L]$ are equivalent representations because

$$\frac{-i}{2}\begin{bmatrix} X & Y - iZ \\ Y + iZ & -X \end{bmatrix} = \begin{bmatrix} 1 & 1 \\ i & -i \end{bmatrix}\frac{-i}{2}\begin{bmatrix} Z & X - iY \\ X + iY & -Z \end{bmatrix}\begin{bmatrix} \frac{1}{2} & -\frac{1}{2}i \\ \frac{1}{2} & \frac{1}{2}i \end{bmatrix}$$

A representation of an algebra will contain a representation of any sub-algebra. If an irreducible representation of an algebra gives a reducible representation of a subalgebra, an equivalent representation may be chosen so that the representation of the subalgebra is a direct sum of irreducible representations. Thus the vectors in a fixed direction form a subalgebra of vector algebra. If the direction is along $\mathbf{i}$, $\mathbf{j}$, or $\mathbf{k}$, the above skew-symmetric representation for the subalgebra is obviously the direct sum of a 2×2 skew-symmetric representation and the 1×1 representation which represents every vector by the number 0. For an arbitrary direction given by the unit vector $\mathbf{l} = l_x\mathbf{i} + l_y\mathbf{j} + l_z\mathbf{k}$, the direct sum is exhibited by an orthogonal transformation

$$T = \begin{bmatrix} l_x & l_y & l_z \\ m_x & m_y & m_z \\ n_x & n_y & n_z \end{bmatrix}$$

with $\mathbf{l} \cdot \mathbf{m} = 0$ and $\mathbf{l} \times \mathbf{m} = \mathbf{n}$. The first column of T^{-1} gives the direction of the subalgebra. Vectors in a plane form a subspace of the vector space, but not a subalgebra, since the cross product is not in the plane.

If the M_A are $n \times n$ matrices, they can be considered as operators acting in the space $\mathcal{V}_n$ of n-dimensional vectors. Suppose that the representation is reducible and the transformation T exhibits the $m \times m$ matrices K_A. Then,

$$M_A T^{-1} = T^{-1} \left[\begin{array}{c|c} K_A & 0 \\ \hline 0 & L_A \end{array} \right]$$

shows that $\mathcal{V}_n$ has an m-dimensional subspace, spanned by the first m columns of T^{-1}, which is invariant under all the matrices M_A. This subspace is called invariant with respect to the representation $[M]$. A representation is reducible if and only if it has such an invariant subspace.

When the A are linear operators, their domain may be represented by the space $\mathcal{V}_n$. For example, consider the representation of o_3 by 3×3 skew-symmetric matrices. Then the representation of the domain associates the vector

$$\begin{bmatrix} a \\ b \\ c \end{bmatrix}$$

of $\mathcal{V}_3$ with the function $ax_1 + bx_2 + cx_3$ (or with $a \cos \theta \cos \phi + b \sin \theta \sin \phi + c \cos \theta$). The matrix representation of o_3 means that this three-dimensional subspace of the domain is invariant under o_3. Because this subspace contains no smaller invariant subspace, the skew-symmetric representation of o_3 is irreducible. Transforming to an equivalent representation will also mean a transformation of the representation of the domain. The transformation

$$T = \begin{bmatrix} -1 & i & 0 \\ 0 & 0 & \sqrt{2} \\ 1 & i & 0 \end{bmatrix}$$

gives the skew-Hermitian representation

$$D_{12} = \begin{bmatrix} -i & 0 & 0 \\ 0 & 0 & 0 \\ 0 & 0 & i \end{bmatrix}, \quad D_{23} = 2^{-1/2} \begin{bmatrix} 0 & -i & 0 \\ -i & 0 & -i \\ 0 & -i & 0 \end{bmatrix},$$

$$D_{31} = 2^{-1/2} \begin{bmatrix} 0 & -1 & 0 \\ 1 & 0 & -1 \\ 0 & 1 & 0 \end{bmatrix} \tag{1.21}$$

and now

$$\begin{bmatrix} 1 \\ 0 \\ 0 \end{bmatrix}, \quad \begin{bmatrix} 0 \\ 1 \\ 0 \end{bmatrix}, \quad \text{and} \quad \begin{bmatrix} 0 \\ 0 \\ 1 \end{bmatrix}$$

respectively represent Y_1^1, Y_1^0, and Y_1^{-1}.

Conversely a matrix representation of an invariant subspace $\mathscr{S}$ of the domain will induce a matrix representation of the operators, which will be irreducible if $\mathscr{S}$ contains no proper subspace which is itself invariant. For example, the function x_i can be represented by the $n \times 1$ column vector with 1 in the ith position and zeros elsewhere $(i = 1, 2, \ldots, n)$. This leads to the representation of $x_i(\partial/\partial x_j)$ by an $n \times n$ matrix with 1 in the (i, j)th position (row i, column j) and zeros elsewhere, and to the $n \times n$ skew-symmetric representation of o_n.

If the space spanned by the $2l + 1$ spherical harmonics is represented by $\mathscr{V}_{2l+1}$, the operators of o_3 are represented by $(2l + 1) \times (2l + 1)$ matrices, and the representation is irreducible. The invariant space $\mathscr{S}_n$ spanned by hyperspherical harmonics leads to an $n^2 \times n^2$ matrix representation of o_4. It is irreducible, but contains reducible representations of the o_3 subalgebras. The correspondence between the hyperspherical harmonics and $\mathscr{V}_{n^2}$ can be chosen to show the representation of one of these subalgebras as a direct sum. For instance if

$$\begin{bmatrix} 1 \\ 0 \\ 0 \\ 0 \end{bmatrix} \leftrightarrow (4\pi R)Z_{2\ 1/2\ 1/2} = -x_1 - ix_2, \qquad \begin{bmatrix} 0 \\ 1 \\ 0 \\ 0 \end{bmatrix} \leftrightarrow (4\pi R)Z_{2\ -1/2\ 1/2} = x_3 - ix_4$$

$$\begin{bmatrix} 0 \\ 0 \\ 1 \\ 0 \end{bmatrix} \leftrightarrow (2\pi R)(Z_{2\ 1/2\ -1/2} + Z_{2\ -1/2\ -1/2}) = \tfrac{1}{2}(x_1 - ix_2 + x_3 + ix_4)$$

$$\begin{bmatrix} 0 \\ 0 \\ 0 \\ 1 \end{bmatrix} \leftrightarrow (2\pi i R)(Z_{2\ -1/2\ -1/2} - Z_{2\ 1/2\ -1/2}) = \tfrac{1}{2}(ix_1 + x_2 - ix_3 + x_4)$$

the matrix in (1.20) represents

$$\tfrac{1}{2}[X(D_{23} + D_{14}) + Y(D_{31} + D_{24}) + Z(D_{12} + D_{34})] = -i(XK_x + YK_y + ZK_z).$$

In general the functions of an invariant space are eigenfunctions of the Casimir operator(s) belonging to the same eigenvalue(s), which may be used to label the representation.

1.6.   REPRESENTATIONS ON A FUNCTION SPACE

Suppose that corresponding to each element A of the algebra $\mathscr{L}$ is an operator A_F defined on a linear space $\mathscr{S}$ of functions f. If this correspondence maps the Lie product in $\mathscr{L}$ onto the commutator, the operators A_F will be referred to as an $\mathscr{L}$ algebra (or subalgebra, as at the end of Section 1.4). The operators and their domain $\mathscr{S}$ are a representation of $\mathscr{L}$ provided $\mathscr{S}$ is invariant under every A_F and the mapping preserves linear relations $(A + B \leftrightarrow A_F + B_F, cA \leftrightarrow cA_F)$.

When the elements of $\mathscr{L}$ are operators, there is usually a mapping between their domain and $\mathscr{S}$ which preserves the action of corresponding operators. For an $n \times n$ matrix representation, the domain $\mathscr{S}$ is understood to be the vector space $\mathscr{V}_n$. If $\mathscr{S}$ is an n-dimensional linear space, to set up the mapping between domains may require either restricting the domain of $\mathscr{L}$ to an n-dimensional invariant subspace or mapping the rest of the domain of $\mathscr{L}$ onto the zero element of $\mathscr{S}$.

For example, the correspondence $D_{23} \leftrightarrow -iK_x$, $D_{31} \leftrightarrow -iK_y$, $D_{12} \leftrightarrow -iK_z$, shows that the components of $-i\mathbf{K}$ form an o_3 algebra. This correspondence together with the statement the domain of $-iK_x$, etc., is the space spanned by the Z_{nfg} with n and g fixed, gives a representation of o_3. A mapping between the domains is $f(r)Y_l^m(\theta, \phi) \rightarrow \delta_{l, \frac{1}{2}n - \frac{1}{2}} Z_{nmg}(\xi, \eta, \phi)$ with f an arbitrary function. Vector algebra gives a representation of o_3 if vectors are regarded as operators acting on vectors by the cross-product rule, so that the domain $\mathscr{S}$ is the same space as the algebra. Suitable correspondences are $D_{23} \leftrightarrow \mathbf{i}$, $D_{31} \leftrightarrow \mathbf{j}$, $D_{12} \leftrightarrow \mathbf{k}$ between operators and $x_1 f(r) \rightarrow \mathbf{i}$, $x_2 f(r) \rightarrow \mathbf{j}$, $x_3 f(r) \rightarrow \mathbf{k}$ between the domains.

Two representations $[F]$ and $[G]$ are equivalent if there is a one-to-one correspondence between their domains such that $f \leftrightarrow g$ implies $(A_F f) \leftrightarrow (A_G g)$ for all A in $\mathscr{L}$. For example, the skew-symmetric matrix representation of o_3 is equivalent to the representation by vector algebra. In the representation of o_3 by components of $-i\mathbf{K}$, taking the $(2l + 1)$ possible values of g gives $(2l + 1)$ equivalent representations. For two equivalent matrix representations the one-to-one correspondence between the domains is given by the transformation T.

If there is no proper, invariant subspace in $\mathscr{S}$, the representation is irreducible. If $\mathscr{S}$ is a direct sum of invariant subspaces $\mathscr{S}_i$, then the representation is reducible and is the direct sum of the representations defined by the domains $\mathscr{S}_i$. Other concepts, such as that of Hermitian representation, require an inner product to be defined on $\mathscr{S}$. If u and v are vectors in $\mathscr{V}_n$, their inner product is u^*v, where u^* is $\bar{u}$ transposed. The various types of matrix characterizing representations can be defined through this inner product; thus M_A is Hermitian if $(M_A u)^*v = u^*M_A v$ for all u and v. This form of the definitions extends to any representation and is familiar in quantum mechanics. An isometric correspondence between domains is one preserving an inner product; representations equivalent through such a correspondence are isometrically equivalent.

The same ideas about representations are used for groups. If A belongs to a group $\mathcal{G}$ and a representation is such that $A_F f$ and f have the same norm for all A and f, the representation is unitary. This implies $A_F^\dagger = A_F^{-1}$. In the next section it will be shown that unitary representations of groups correspond to skew-Hermitian representations of algebras. The latter only exist for real algebras. For calculations it may be convenient to use the complex extension (in particular, (1.12) is then available), but the results obtained are interpreted in terms of the real algebra.

From the mathematical viewpoint, an algebra is an abstract entity defined by the structure constants which give the Lie products of a basis. For instance (1.8) can define o_n, and (1.7) then becomes a representation. Here (1.7) has been used as the definition, and referred to as *the* algebra o_n, because it is naturally related to the rotation group, which is both familiar and relevant to most applications. The relation to the group is discussed in the next section.

Quantum mechanics considers kets corresponding to the states of a physical system and for calculations uses representations of the kets, usually functions of position, functions of momentum, or vectors from $\mathcal{V}_n$. This is quite consistent with the notions of a representation given above, but to avoid overworking this word wave functions (of position or momentum) will be referred to as realizations of the kets. In this book, then, representation always implies an algebra or a group.

1.7. ROTATION GROUPS

The two-dimensional rotation group is the set of operators $R(t)$ defined by $R(t)f = g$, $g(\phi) = f(\phi + t)$, with domain consisting of analytic functions f of period $2\pi : f(\phi) = f(\phi + 2\pi)$. From Taylor's theorem

$$f(\phi + t) = \sum_{n=0}^{\infty} \frac{t^n}{n!} f^{(n)}(\phi) = \sum_{n=0}^{\infty} \frac{1}{n!} \left(t\frac{\partial}{\partial \phi} \right)^n f(\phi) = \exp\left(t\frac{\partial}{\partial \phi} \right) f(\phi)$$

assuming that the exponential series can define the exponential function of an operator. Thus the group consists of the operators $\exp(t(\partial/\partial\phi))$ with t real. If ϕ is the polar angle in a plane, then $\partial/\partial\phi = D_{21}$, and the operators $t(\partial/\partial\phi)$ are the algebra o_2. So the group may be obtained by considering the exponential function on an algebra.

The other groups needed for this book will now be introduced in this way. An exponential function e^A may be defined on an algebra of operators A. The exponential theorem $e^A e^B = e^{A+B}$ is true when $AB = BA$, but in general $e^A e^B = e^C$, where C can be expressed (J62) in terms of A, B, and the commutators that can be formed from A and B, for example, $[[A, [A, B]], B]$. Then if the algebra is a Lie algebra, C is a member of the Lie algebra, and the operators e^A form a group.

When the elements of the algebra are not operators, the idea can still be applied to the matrix representations. Thus for vector algebra, $a\mathbf{i} + b\mathbf{j} + c\mathbf{k}$ is represented by

$$M = \begin{bmatrix} 0 & -c & b \\ c & 0 & -a \\ -b & a & 0 \end{bmatrix}$$

If

$$P = \begin{bmatrix} a \\ b \\ c \end{bmatrix} \begin{bmatrix} a & b & c \end{bmatrix}$$

and $\theta^2 = a^2 + b^2 + c^2$, then $M^2 = P - \theta^2 I$, $MP = 0$, leading (by induction) to $M^{2n} = (-\theta^2)^n(I - \theta^{-2}P)$, $M^{2n+1} = (-\theta^2)^n M$. So the exponential series gives

$$e^M = (\cos \theta)I + (\theta^{-1} \sin \theta)M + \theta^{-2}(1 - \cos \theta)P \qquad (1.22)$$

Putting $N = \theta^{-1}M$ and $Q = \theta^{-2}P$, which are matrices corresponding to unit vectors, gives

$$e^{\theta N} = (\cos \theta)I + (\sin \theta)N + (1 - \cos \theta)Q \qquad (1.23)$$

The transpose $\tilde{M}$ of M is $-M$, so transposing the exponential series shows that $\widetilde{e^M} = e^{-M}$. Thus e^M is orthogonal. Also a general property of the exponential function of a matrix is

$$|e^A| = \exp(\operatorname{tr} A) \qquad (1.24)$$

where $\operatorname{tr} A$ is the trace of A (the sum of the diagonal elements). Since $\operatorname{tr} M = 0$, e^M is unimodular and is thus a proper orthogonal matrix. The e^M are therefore a matrix representation of a group of rotations in three dimensions. In fact the right side of (1.23) is a matrix representation of a rotation through angle θ about the unit vector $\mathbf{n}$ represented by N. Let a fixed point have coordinates (x, y, z) relative to $Oxyz$, and coordinates (x', y', z') relative to axes obtained by rotating $Oxyz$ through angle θ about $\mathbf{n}$. Then (S60)

$$[x' \ y' \ z'] = [x \ y \ z]e^{\theta N} \qquad (1.25)$$

This also shows that the e^M give all proper orthogonal matrices, since all possible rotations are obtained.

The same procedure can be applied to the algebra o_3. If A belongs to o_3, then e^A is a rotation operator which acts on functions of three variables by an orthogonal coordinate transformation corresponding to a rotation of axes. For example, $\exp(\theta D_{12})f = g$, $g(x, y, z) = f(x', y', z')$ with

$$x' = x \cos \theta + y \sin \theta, \; y' = -x \sin \theta + y \cos \theta, \; z' = z \qquad (1.26)$$

For $A = aD_{23} + bD_{31} + cD_{12}$, the transformation is given by (1.25) and (1.23), with $\theta N = M$ the same 3×3 skew-symmetric matrix. The resulting group SO_3 of transformation operators is the three-dimensional rotation group. Similarly the group obtained from the algebra o_n is the group of transformation operators SO_n corresponding to proper orthogonal transformations of $x_1, x_2, \ldots, x_n$. These coordinate transformations leave the expression $x_1{}^2 + x_2{}^2 + \cdots + x_n{}^2$ invariant and are obtained geometrically by rotating the coordinate axes. For any operator A of the algebra the transformation matrix is e^M, where M represents A in the $n \times n$ skew-symmetric representation.

A basis for the Lie algebra is called a set of generators of the corresponding group. When the algebra operators A are skew-Hermitian, it is usual in quantum mechanics to refer to the Hermitian operators iA as the generators. For example the operators appearing in (1.9) and (1.10) are called generators of the three- and four-dimensional rotation groups.

The exponential function defined on a matrix representation of the algebra will give a matrix representation of the group. A set of functions that is invariant under the group will also be invariant under the algebra, so that irreducible representations of the algebra give irreducible representations of the group. Thus the $2l + 1$ spherical harmonics $Y_l{}^m$ are invariant under rotations. The hyperspherical harmonics Y_{nlm} are invariant under the group SO_4 of four-dimensional rotations, giving a representation of this group of dimension n^2. The o_3 subalgebra spanned by D_{12}, D_{23}, and D_{31} corresponds to the subgroup SO_3 of rotations in the first three dimensions, and the representation of SO_3 thus obtained is the direct sum of the irreducible representations with $l = 1, 2, \ldots, n - 1$.

An element of the algebra o_3, which has dimension 3, can be specified by giving three real parameters, and the same is true for the group SO_3. The parameters for the algebra could be the components of the isomorphic vector $\theta \mathbf{n}$, and the same parameters used for the group with the restriction $\theta \le \pi$. Matrix representations of the group are usually given in terms of Euler angles α, β, γ as parameters. The general group element is then

$$R = \exp(\alpha D_{12}) \exp(\beta D_{31}) \exp(\gamma D_{12})$$

where the parameters can be restricted to $0 \le \alpha < 2\pi$, $0 \le \gamma < 2\pi$, $0 \le \beta < \pi$. In quantum mechanics the representations of interest are those giving a Hermitian representation of L_x, L_y, L_z, hence a skew-Hermitian representation of the $D_{\alpha\beta}$, for example (1.21). The representation of the group is then unitary. In the representation $[l]$ obtained from the invariant subspace having the spherical harmonics Y_l^m as basis, R is represented by a matrix with elements $(m, n = -l, \ldots, +l) D_{mn}^l(\alpha, \beta, \gamma) = d_{mn}^l(\beta) \exp(-i\alpha m - i\gamma n)$ with d_{mn}^l given by (1.18):

$$RY_l^n = \sum_m Y_l^m D_{mn}^l \qquad (1.27)$$

1.8.   ADDITION THEOREMS FOR SPHERICAL AND HYPERSPHERICAL HARMONICS

Equation (1.27) is the expansion in spherical harmonics of the function RY_l^n. On the line with spherical polar angles (θ, ϕ) the value of this function is $Y_l^n(\theta', \phi')$, where θ' and ϕ' are coordinates relative to rotated axes and are given by (1.25). Equation (1.27) can therefore also be regarded as giving the relation between the values of spherical harmonics evaluated in different polar coordinate systems. This interpretation will be used in this section.

The addition theorem for spherical harmonics states that if $\Omega_1 = (\theta_1, \phi_1)$ and $\Omega_2 = (\theta_2, \phi_2)$ are the polar angles of two vectors $\mathbf{r}_1$ and $\mathbf{r}_2$ and γ is the angle between them, then

$$4\pi \sum_m Y_l^m(\Omega_1) \overline{Y}_l^m(\Omega_2) = (2l + 1) P_l(\cos \gamma) \qquad (1.28)$$

The angle γ, and hence the right side of the equation, does not depend on the coordinate system used to define the polar angles of the vectors. The implied invariance of the left side under rotations of the axes leads to the following proof (R57) of the result.

If Ω_i' are coordinates relative to rotated axes, then

$$\sum_n Y_l^n(\Omega_1') \overline{Y}_l^n(\Omega_2') = \sum_{nmk} Y_l^m(\Omega_1) D_{mn}^l \overline{D}_{kn}^l \overline{Y}_l^k(\Omega_2) \qquad \text{using (1.27)}$$

$$= \sum_{mk} Y_l^m(\Omega_1) \delta_{m,k} \overline{Y}_l^k(\Omega_2)$$

since $[l]$ is unitary. Thus the left side of (1.28) is invariant under rotations in the sense that its value is the same function of the four polar angles in any coordinate system. It can therefore be evaluated using any convenient rotated axes. Take the polar axis along $\mathbf{r}_2$, so that $\theta_1' = \gamma$ and $\theta_2' = 0$. As ϕ is irrelevant on the polar axis, $Y_l^m(0, \phi)$ must be independent of ϕ; but Y_l^m depends on ϕ through the factor $e^{im\phi}$, so $Y_l^m(0, \phi)$ must be zero if $m \neq 0$. Then $\sum_n Y_l^n(\Omega_1') \overline{Y}_l^n(\Omega_2') = Y_l^0(\gamma, \phi_1') \overline{Y}_l^0(0, \phi_2')$. From (1.13) and the Rodrigues formula (H65) for the Legendre polynomials,

$$Y_l^0(\theta, \phi) = \tfrac{1}{2}(2l + 1)^{1/2} \pi^{-1/2} P_l(\cos \theta).$$

Thus (1.28) is established, since (H65) $P_l(1) = 1$.

Another result easily demonstrated from this interpretation of (1.27) is the unitary property of the representation. From (1.25) the Jacobian of the coordinate transformation is $|e^{\theta N}| = 1$, so that $d\Omega = d\Omega'$. Hence

$$\delta_{m,k} = \int \overline{Y}_l^m(\Omega')Y_l^k(\Omega')\, d\Omega'$$

$$= \sum_{np} \overline{D}_{nm}^l D_{pk}^l \int \overline{Y}_l^n(\Omega)Y_l^p(\Omega)\, d\Omega = \sum_n \overline{D}_{nm}^l D_{nk}^l$$

This proof extends to give an addition theorem for the hyperspherical harmonics. If $\Delta_{lm,\ LM}^n$ denotes the elements of the $n^2 \times n^2$ matrix giving an irreducible representation of a rotation in four dimensions, then the analog of (1.27) is

$$Y_{nLM}(\Omega') = \sum_{lm} Y_{nlm}(\Omega)\ \Delta_{lm,\ LM}^n$$

in which Ω (and Ω') now stand for three angles α, θ, ϕ. Because the matrix Δ^n is unitary, $\sum_{lm} Y_{nlm}(\Omega_1)\,\overline{Y}_{nlm}(\Omega_2)$ can be evaluated in any suitable coordinate system. Choosing the 4-axis in the Ω_2 direction, $\alpha_2 = 0$, and $\alpha_1 = \gamma$, the angle between the two directions. Since θ and ϕ are irrelevant when $\gamma = 0$, $Y_{nlm}(0, \theta, \phi)$ must be independent of θ and ϕ, hence zero unless $l = m = 0$. From (1.15), $2^{1/2}\pi Y_{n00}(\alpha, \theta, \phi) = i^{n-1}C_{n-1}^1(\cos \alpha)$, and (H65) $C_{n-1}^1(1) = n$. Combining these results gives the addition theorem

$$2\pi^2 \sum_{lm} Y_{nlm}(\alpha_1, \theta_1, \phi_1)\,\overline{Y}_{nlm}(\alpha_2, \theta_2, \phi_2) = nC_{n-1}^1(\cos \gamma) \qquad (1.29)$$

in which γ is the angle between the two directions (in four-dimensional space) defined by the angles $(\alpha_1, \theta_1, \phi_1)$ and $(\alpha_2, \theta_2, \phi_2)$. An alternative proof of the result is given by Talman (T68); it can also be derived by first using (1.28) and then an addition theorem (M43–54) for the Gegenbauer polynomials.

Since the hyperspherical harmonics Y_{nlm} of (1.15) and Z_{nfg} of (1.17) are connected by a unitary transformation, the result (1.29) is also true for the Z_{nfg}:

$$2\pi^2 \sum_{fg} Z_{nfg}(\xi_1, \eta_1, \phi_1)\overline{Z}_{nfg}(\xi_2, \eta_2, \phi_2) = nC_{n-1}^1(\cos \gamma)$$

This can be derived independently by the same method: evaluate the left side with the 1-axis in the (ξ_1, η_1, ϕ_1) direction and the 2-axis in the plane containing both directions, so that $\xi_1 = \phi_1 = \xi_2 = 0$, and $\phi_2 = \gamma$. From (1.18), $d_{fg}^F(0) = \delta_{f,g}$, and the left side is a geometric progression:

$$n \sum_{f=-F}^{F} e^{2if\gamma} = \frac{ne^{-2iF\gamma}(e^{2i\gamma n} - 1)}{e^{2i\gamma} - 1} \qquad (n = 2F + 1)$$

$$= \frac{n(e^{in\gamma} - e^{-in\gamma})}{e^{i\gamma} - e^{-i\gamma}} = \frac{n \sin n\gamma}{\sin \gamma} = nC_{n-1}^1(\cos \gamma)$$

1.9. PSEUDOROTATION GROUPS

The n-dimensional rotation group SO_n is a set of transformation operators defined by coordinate transformations which leave $x_1{}^2 + x_2{}^2 + \cdots + x_n{}^2$ invariant. The pseudorotation groups are obtained by considering the transformations which leave $x_1{}^2 + x_2{}^2 + \cdots + x_p{}^2 - x_{p+1}^2 - \cdots - x_n{}^2$ invariant. With each group is associated a real Lie algebra $o_{p,\,n-p}$ and the group may be obtained from an exponential function defined on the algebra.

The simplest example is $o_{2,1}$ with the basis

$$D_{12} = -x_1 \frac{\partial}{\partial x_2} + x_2 \frac{\partial}{\partial x_1}, \quad E_{23} = x_2 \frac{\partial}{\partial x_3} + x_3 \frac{\partial}{\partial x_2},$$

$$E_{31} = x_3 \frac{\partial}{\partial x_1} + x_1 \frac{\partial}{\partial x_3} \tag{1.30}$$

such that

$$[D_{12}, E_{23}] = -E_{31}, [E_{23}, E_{31}] = D_{12}, [E_{31}, D_{12}] = -E_{23} \tag{1.31}$$

The correspondence $E_{23} \leftrightarrow L_z$, $E_{31} + D_{12} \leftrightarrow L_+$, $E_{31} - D_{12} \leftrightarrow L_-$ shows that another $o_{2,1}$ algebra is the real Lie algebra mentioned in Section 1.2 defined by the products $[L_z, L_\pm] = \pm L_\pm$ and $[L_+, L_-] = 2L_z$. However, this does not mean that angular momentum theory gives all the properties of $o_{2,1}$, because E_{23} is not Hermitian. On the other hand, the complexification of $o_{2,1}$ is a $c * o_3$ algebra, the natural mapping being $-iE_{23} \leftrightarrow D_{23}$, $iE_{31} \leftrightarrow D_{31}$; it contains $iE_{23} \pm E_{31}$ which are raising and lowering operators for the integer eigenvalues of iD_{12}. The Casimir operator is $D_{12}^2 - E_{23}^2 - E_{31}^2$.

The previous matrix representation of $x_i(\partial/\partial x_j)$ immediately gives the 3×3 representation of $o_{2,1}$, but E_{23} and E_{31} are represented by symmetric matrices, and so an exponential function does not give orthogonal matrices. (1.22) is valid with

$$M = \begin{bmatrix} 0 & -c & b \\ c & 0 & a \\ b & a & 0 \end{bmatrix}, \quad P = \begin{bmatrix} -a \\ b \\ c \end{bmatrix} \begin{bmatrix} a & -b & c \end{bmatrix}$$

and $\theta^2 = c^2 - a^2 - b^2$, so that again $MP = PM = 0$, $P^2 = \theta^2 P$ and $M^2 = P - \theta^2 I$. If

$$I^+ = \begin{bmatrix} 1 & 0 & 0 \\ 0 & 1 & 0 \\ 0 & 0 & -1 \end{bmatrix}$$

then $\tilde{M}I^+ = -I^+M$, $\tilde{P}I^+ = I^+P$, and (1.22) gives $\widetilde{e^M}I^+e^M = I^+$, which says that the exponential function gives matrices of transformations leaving

$x_1{}^2 + x_2{}^2 - x_3{}^2$ invariant. The operator $\exp(aE_{23} + bE_{31} + cD_{12})$ acts on a function according to the coordinate transformation with matrix e^M. The general group element can be written $\exp(\alpha D_{12}) \exp(\beta E_{23}) \exp(\gamma D_{12})$ and the matrix of the corresponding transformation is

$$
\begin{bmatrix} \cos\alpha & -\sin\alpha & 0 \\ \sin\alpha & \cos\alpha & 0 \\ 0 & 0 & 1 \end{bmatrix}
\begin{bmatrix} 1 & 0 & 0 \\ 0 & \cosh\beta & \sinh\beta \\ 0 & \sinh\beta & \cosh\beta \end{bmatrix}
\begin{bmatrix} \cos\gamma & -\sin\gamma & 0 \\ \sin\gamma & \cos\gamma & 0 \\ 0 & 0 & 1 \end{bmatrix}
$$

$$(1.32)$$

This has determinant 1, and the element in the third row and column is $\cosh\beta \geq 1$. The group represented by such matrices will be denoted by $SO_{2,1}$.

An important difference between the groups SO_3 and $SO_{2,1}$ is now apparent. All the elements of SO_3 are obtained as the Euler angle β varies over the *bounded* interval $0 \leq \beta < \pi$, but to get all elements of $SO_{2,1}$ the analogous parameter β must vary over the *unbounded* interval $\beta \geq 0$. The group SO_3 is said to be compact while $SO_{2,1}$ is noncompact. The generator E_{23} which gets multiplied by β is called a noncompact generator. If a representation of a noncompact group is unitary and irreducible, then the dimension of the representation is either 1×1 or not finite. The representations used in quantum mechanics will therefore be infinite dimensional.

The real Lie algebra $o_{p,\,n-p}$ is of dimension $\frac{1}{2}n(n-1)$ and is spanned by the operators ($D_{\beta\alpha} = -D_{\alpha\beta}$, $E_{\beta\alpha} = E_{\alpha\beta}$):

$$
D_{\alpha\beta} = -x_\alpha \frac{\partial}{\partial x_\beta} + x_\beta \frac{\partial}{\partial x_\alpha} \quad (1 \leq \alpha < \beta \leq p, \, p < \alpha < \beta \leq n)
$$

$$(1.33)$$

$$
E_{\alpha\beta} = x_\alpha \frac{\partial}{\partial x_\beta} + x_\beta \frac{\partial}{\partial x_\alpha} \quad (1 \leq \alpha \leq p, \, p < \beta \leq n)
$$

Operators with no common subscript commute, and otherwise

$$
[D_{\alpha\beta}, D_{\beta\gamma}] = D_{\gamma\alpha}, \; [D_{\alpha\beta}, E_{\beta\gamma}] = -E_{\alpha\gamma}, \; [E_{\alpha\beta}, E_{\beta\gamma}] = D_{\gamma\alpha} \qquad (1.34)
$$

The functions $x_1, x_2, \ldots, x_n$ are a basis for an n-dimensional invariant space, inducing an irreducible $n \times n$ representation in which each matrix M has the partitioned form $\begin{bmatrix} A & B \\ \tilde{B} & C \end{bmatrix}$ where A is $p \times p$ and skew-symmetric and C is $(n-p) \times (n-p)$ and skew-symmetric. The matrices e^M are a group of unimodular matrices of transformations leaving $x_1{}^2 + \cdots + x_p{}^2 - \cdots - x_n{}^2$ invariant. The corresponding group of transformation operators will be called $SO_{p,\,n-p}$. It is sufficient to consider $p \geq n - p$, as the groups $SO_{p,q}$ and $SO_{q,p}$ are isomorphic. The case $p = n$ gives the rotation groups, which are compact, and for which the $n \times n$ representation is orthogonal. The groups with $p < n$ are called pseudorotation groups and are not compact. The proper Lorentz group is $SO_{3,1}$, while the groups $SO_{4,1}$ and $SO_{3,2}$ are called de Sitter groups.

In Appendix D the commutators (1.34) for the algebras $o_{3,2}$ and $o_{4,2}$ are tabulated. The skew-Hermitian operators (1.33) are changed to the Hermitian operators appearing in the physical problems, with the notations $(L_1, L_2, L_3) = (iD_{23}, iD_{31}, iD_{12})$; $V_\alpha = iE_{\alpha 5}(\alpha = 1, 2, 3)$; for $o_{3,2}$, $U_\alpha = iE_{\alpha 4}(\alpha = 1, 2, 3)$ and $S = iD_{45}$; for $o_{4,2}$, $V_4 = iE_{45}$, $M_\alpha = iD_{\alpha 4}$ $(\alpha = 1, 2, 3)$, $W_\alpha = iE_{\alpha 6}$ $(\alpha = 1, 2, 3, 4)$ and $N = iD_{65}$. The first six columns of Table 2 give (1.10), the first ten columns describe a basis of the de Sitter subalgebra $o_{4,1}$, while the first six columns of Table 1 involve the subalgebra $o_{3,1}$.

1.10. THE $o_{3,2}$ ALGEBRA ASSOCIATED WITH SPHERICAL HARMONICS

The spherical harmonics defined in (1.13) are eigenfunctions of L_z, and $L_\pm = L_x \pm iL_y$ are shift (raising and lowering) operators for L_z:

$$L_z Y_l^m = m Y_l^m, \quad L_\pm Y_l^m = (l \mp m)^{1/2}(l \pm m + 1)^{1/2} Y_l^{m \pm 1} \qquad (1.35)$$

These equations, which lead to a representation of o_3, could be used to define L_z and $L_\pm$. Shift operators for l may be defined in this way as follows:

$$A_0 Y_l^m = (l^2 - m^2)^{1/2} Y_{l-1}^m, \quad A_\pm Y_l^m = (l \pm m)^{1/2}(l \pm m - 1)^{1/2} Y_{l-1}^{m \mp 1} \qquad (1.36)$$

Then as $A_0^\dagger Y_{l-1}^m = (l^2 - m^2)^{1/2} Y_l^m$, etc., the complex conjugates satisfy

$$A_0^\dagger Y_l^m = (l + 1 + m)^{1/2}(l + 1 - m)^{1/2} Y_{l+1}^m$$
$$A_\pm^\dagger Y_l^m = (l \pm m + 1)^{1/2}(l \pm m + 2)^{1/2} Y_{l+1}^{m \pm 1} \qquad (1.37)$$

The commutators of all these operators can be worked out from these definitions. For example,

$$A_+ A_+^\dagger Y_l^m = (l + m + 1)^{1/2}(l + m + 2)^{1/2} A_+ Y_{l+1}^{m+1}$$
$$= (l + m + 1)(l + m + 2) Y_l^m$$

and

$$A_+^\dagger A_+ Y_l^m = (l + m)(l + m - 1) Y_l^m$$
$$[A_+, A_+^\dagger] Y_l^m = (4l + 4m + 2) Y_l^m.$$

Defining S by $SY_l^m = (l + \tfrac{1}{2}) Y_l^m$, $[A_+, A_+^\dagger] = 4S + 4L_z$. No further operators are needed to express all commutators as linear combinations of the ten operators $L_z, L_\pm, A_k^\dagger, A_k, S(k = 0, \pm)$, which therefore form a Lie algebra.

Taking this algebra over the complex field, the subset of skew-Hermitian operators form a real Lie algebra. This can be identified with $o_{3,2}$ by choosing the following basis and comparing its commutators with Table 1 of Appendix D.

$$D_{12} \leftrightarrow -iL_z, D_{23} \leftrightarrow -\tfrac{1}{2}i(L_+ + L_-), D_{31} \leftrightarrow \tfrac{1}{2}(L_- - L_+), D_{45} \leftrightarrow -iS,$$

$$E_{14} \leftrightarrow \tfrac{1}{4}i(A_-^\dagger + A_- - A_+^\dagger - A_+), E_{24} \leftrightarrow \tfrac{1}{4}(A_+ - A_+^\dagger + A_- - A_-^\dagger)$$

$$E_{34} \leftrightarrow \tfrac{1}{2}i(A_0^\dagger + A_0), E_{15} \leftrightarrow \tfrac{1}{4}(A_-^\dagger - A_- - A_+^\dagger + A_+) \tag{1.38}$$

$$E_{25} \leftrightarrow \tfrac{1}{4}i(A_+^\dagger + A_+ + A_-^\dagger + A_-), E_{35} \leftrightarrow \tfrac{1}{2}(A_0^\dagger - A_0)$$

Differential operators for raising and lowering l were given by Infeld and Hull (I51); for instance,

$$A_0 = (2l - 1)^{1/2}(2l + 1)^{-1/2}\left(l \cos \theta - \sin \theta \, \frac{\partial}{\partial \theta}\right)$$

If such expressions are used to evaluate commutators or complex conjugates it is important to remember that l is really an operator rather than a number. Using S defined above gives

$$A_0 = \left\{(\cos \theta)\left(S - \frac{1}{2}\right) - \sin \theta \, \frac{\partial}{\partial \theta}\right\}(1 - S^{-1})^{1/2}$$

The operators of the $o_{3,2}$ algebra are linearly independent but not algebraically independent. Obviously $S^2 = L^2 + \tfrac{1}{4}$, and other identities, such as $A_0 A_0^\dagger = (S + \tfrac{1}{2})^2 - L_z^2$, appear during the calculation of commutators.

The work in this section may be summarized as follows: the operators (1.38), defined on the spherical harmonics by the rules (1.35), (1.36), and (1.37), are a representation of $o_{3,2}$. The domain, or representation space, is the set of functions $f(\theta, \phi)$ linearly dependent on the spherical harmonics. With the usual inner product ($\|f\|^2 = \iint |f(\theta, \phi)|^2 \sin \theta \, d\theta \, d\phi$) this representation is skew-Hermitian.

1.11.   THE GROUPS SU_2, U_2, AND $SU_{1,1}$

The general element A of the algebra o_3 can be written

$$A = \theta(n_1 D_{23} + n_2 D_{31} + n_3 D_{12})$$

with $n_1^2 + n_2^2 + n_3^2 = 1$. In (1.20) there is a 2×2 matrix representation of o_3 in which A is represented by $K = -\tfrac{1}{2}i\theta(n_1\sigma_1 + n_2\sigma_2 + n_3\sigma_3)$, where σ_1, σ_2, and σ_3 are the Pauli spin matrices with the properties

$$\sigma_1^2 = \sigma_2^2 = \sigma_3^2 = I, \sigma_1\sigma_2 = -\sigma_2\sigma_1 = i\sigma_3, \cdots \tag{1.39}$$

Then $K^{2n} = (K^2)^n = (-)^n(\tfrac{1}{2}\theta)^{2n}I$, and

$$K^{2n+1} = -i(-)^n(\tfrac{1}{2}\theta)^{2n+1}(n_1\sigma_1 + n_2\sigma_2 + n_3\sigma_3).$$

From the exponential series

$$e^K = (\cos\tfrac{1}{2}\theta)I - i(\sin\tfrac{1}{2}\theta)(n_1\sigma_1 + n_2\sigma_2 + n_3\sigma_3) = \begin{bmatrix} \xi & \eta \\ -\bar\eta & \bar\xi \end{bmatrix} \quad (1.40)$$

where $\xi = \cos\tfrac{1}{2}\theta - in_3\sin\tfrac{1}{2}\theta$, $\eta = -(n_2 + in_1)\sin\tfrac{1}{2}\theta$. Then $|\xi|^2 + |\eta|^2 = 1$, so it can be verified directly that the matrix e^K is unitary and has determinant 1. This can also be deduced as in Section 1.7, since $K = -\tilde{K}$ and $\operatorname{tr} K = 0$.

This representation $[K]$ of o_3, called the fundamental representation of the algebra, consists of the set of 2×2 skew-Hermitian matrices with zero trace. The exponential function of these matrices gives the group of 2×2 unitary, unimodular matrices, which is the fundamental representation of the group SU_2 of operators acting on functions $f(z, w)$ of two complex variables by unitary, unimodular transformations of the variables. Linear functions $L(z, w) = az + bw$ form an invariant space. If $\begin{bmatrix} a \\ b \end{bmatrix}$ represents L, then (1.40) represents the operator which changes f to $f(\xi z - \bar\eta w, \eta z + \bar\xi w)$. The group can also be defined as the transformations leaving $z\bar z + w\bar w$ invariant.

Both the groups SO_3 and SU_2 have now been obtained by taking an exponential function on a representation of o_3. The mapping of SU_2 onto SO_3 in which (1.23) corresponds to (1.40) is a homomorphism since multiplication is preserved. It is not an isomorphism, because (1.40) changes sign on replacing θ by $2\pi - \theta$ and n_i by $-n_i$, whereas (1.23) is invariant under these substitutions. All proper rotations are obtained from (1.23) by taking $0 \le \theta \le \pi$, but $0 \le \theta \le 2\pi$ is required in (1.40) to get all the unimodular, unitary matrices. There are two such matrices corresponding to each rotation, and therefore (1.40) does not give a matrix representation of SO_3. It is impossible to assign one of the unitary matrices to each rotation and still preserve all multiplications. On the other hand, the 3×3 matrices (1.23) are a representation of the group SU_2.

Therefore the association of a group with an algebra by means of an exponential function does not lead to a unique group. In the standard theory of Lie groups (T68), the algebra is obtained from the group, and the process is unique. In applications to quantum mechanics, the existence of an algebra is often more obvious than the existence of the corresponding group. This is especially true of the groups and algebras to be used later in this book. If an algebra is identified, the nature of the relevant group still has to be discussed. The groups SU_2 and SO_3 corresponding to o_3 are easily distinguished by the dimensions of the invariant subspaces of the algebra, that is, by the dimensions of the obtained irreducible representations of the algebra. The SU_2

group has irreducible representations of every dimension, while the SO_3 group has irreducible representations of odd dimension only, their domains being spanned by spherical harmonics Y_l^m ($m = -l, \ldots, l$). So if an o_3 algebra has irreducible representations of even dimension the corresponding group will be SU_2.

For o_3 an equivalent way of identifying the group is from the eigenvalues of the generators. In the representations of o_3 of odd order, the eigenvalues of the operators representing $L_x = iD_{23}$, $L_y = iD_{31}$, or $L_z = iD_{12}$ are integers. In the representations of even order, these operators will have half-integral eigenvalues. The appearance of half-integral eigenvalues therefore indicates the SU_2 group rather than SO_3. The restriction to integers of the eigenvalues of the generators of the rotation group follows by considering the invariance of any function under a rotation through 2π. If k is an integer, replacing θ by $\theta + 2k\pi$ in (1.26) gives the same transformation. An eigenfunction of L_z belonging to the eigenvalue m is an eigenfunction of the transformation $\exp(\theta D_{12})$ belonging to the eigenvalue $e^{-im\theta}$. Then $e^{-im(\theta + 2k\pi)}$ must be the same number for any integer k, which requires m to be an integer. However, the transformation (1.40) is only invariant when θ is changed by $4k\pi$, and so the same argument restricts m to be integral or half-integral.

The algebra o_3 and the group SU_2 have irreducible matrix representations of every order, and two representations of the same order are equivalent. An $n \times n$ representation is obtained from the invariant space consisting of all homogeneous polynomials of degree $n-1$ in z and w. The operation $f(z, w) \to f(-z, -w)$ multiplies a homogeneous polynomial by $(-)^{n-1}$, and is therefore represented by $(-)^{n-1}I$.

The group U_2 of unitary transformations has representations which are simply related to those of SU_2. The fundamental representation of U_2 is just the group of 2×2 unitary matrices U, which have determinant $e^{i\gamma}$ with γ real. Then $S = e^{-\frac{1}{2}i\gamma}U$ is unimodular. If S_M represents S in an $n \times n$ representation of SU_2, the correspondence $U \to e^{\frac{1}{2}i\gamma r}S_M$ preserves multiplication for any number r. However, r must be restricted because U can be expressed as a product of a unimodular unitary matrix and a number in two ways:

$$U = e^{\frac{1}{2}i\gamma}S = e^{\frac{1}{2}i\gamma + i\pi}(-S)$$

Now $-S = \begin{bmatrix} -1 & 0 \\ 0 & -1 \end{bmatrix} S$ is represented by $(-)^{n-1}IS_M = (-)^{n-1}S_M$, so the correspondence associates with U both the matrices $e^{\frac{1}{2}i\gamma r}S_M$ and $e^{\frac{1}{2}i(\gamma + 2\pi)r}(-)^{n-1}S_M$. These are the same if $(-)^{n+r-1} = 1$, so r must be an integer with the same parity as $n-1$.

Thus an infinity of representations of U_2 may be obtained from an irreducible representation of SU_2. These representations of U_2 are not equivalent:

if $Te^{\frac{1}{2}i\gamma r}S_M T^{-1} = e^{\frac{1}{2}i\gamma s}S_M$, then $\gamma = 0$ gives $TS_M T^{-1} = S_M$, $S_M T^{-1} = T^{-1}S_M$, and so $e^{\frac{1}{2}i\gamma r}S_M = e^{\frac{1}{2}i\gamma s}S_M$ (for all γ), which is only possible if $r = s$.

In Section 1.9 the group $SO_{2,1}$ was obtained from the 3×3 representation of the algebra $o_{2,1}$, by analogy with the treatment of o_3 and SO_3 in Section 1.7. The work in this section leading from o_3 to the group SU_2 also has its parallel for $o_{2,1}$. The fundamental representation of this algebra is $D_{12} \leftrightarrow -\frac{1}{2}i\sigma_3$, $E_{23} \leftrightarrow \frac{1}{2}\sigma_1$, $E_{31} \leftrightarrow -\frac{1}{2}\sigma_2$; the representation consists of all 2×2 matrices of the form $K = \begin{bmatrix} -i\varepsilon & \zeta \\ \bar{\zeta} & i\varepsilon \end{bmatrix}$ with ε an arbitrary real number and ζ an arbitrary complex number. If θ is defined by $\theta^2 = 4\varepsilon^2 - 4\zeta\bar{\zeta}$, then $K^2 = -\frac{1}{4}\theta^2 I$, and the exponential series for e^K can again be summed by taking even and odd powers separately. If $\theta = 0$, then $e^K = I + K$; if $\theta \neq 0$, then

$$e^K = (\cos \tfrac{1}{2}\theta)I + 2\theta^{-1}(\sin \tfrac{1}{2}\theta)K,$$

which is merely (1.40) rewritten. For both the o_3 and $o_{2,1}$ representations, $\theta^2 = 4|K|$. However, whereas for o_3 $|K| > 0$ if $K \neq 0$, and θ is real, for $o_{2,1}$ the determinant can have either sign and can be zero for $K \neq 0$. Thus for $o_{2,1}$ θ can be real or pure imaginary; however, $\cos \frac{1}{2}\theta$ and $2\theta^{-1}(\sin \frac{1}{2}\theta)$ are always real. Hence

$$e^K = \begin{bmatrix} \xi & \eta \\ \bar{\eta} & \bar{\xi} \end{bmatrix} \quad \text{with} \quad \begin{cases} \xi = \cos \tfrac{1}{2}\theta - 2i\theta^{-1}\varepsilon \sin \tfrac{1}{2}\theta \\ \eta = 2\zeta\theta^{-1} \sin \tfrac{1}{2}\theta \end{cases} \tag{1.41}$$

Since $|\xi|^2 - |\eta|^2 = 1$, the corresponding transformations

$$f \to f(\xi z + \bar{\eta}w, \eta z + \bar{\xi}w)$$

leave $z\bar{z} - w\bar{w}$ invariant. The group $SU_{1,1}$ of such transformations is pseudo-unitary, but it may still have unitary representations corresponding to skew-Hermitian representations of the algebra.

In the $SU_{1,1}$ transformations, η is an arbitrary complex number, and to generate all such transformations from the algebra requires ζ to take all complex values. This is clear from the transformations with ξ real, for then $\varepsilon = 0$, $\xi = \cosh |\zeta|$, and $\eta = (\sinh |\zeta|) \arg \zeta$. The group $SU_{1,1}$ is therefore noncompact. In K the real and imaginary parts of $\frac{1}{2}\zeta$ are the coefficients of the matrices representing E_{23} and E_{31}, which are thus noncompact generators. The noncompactness of $SU_{1,1}$ also follows from the analog of (1.32): the general element of the fundamental representation can be written

$$\begin{bmatrix} e^{\frac{1}{2}i\alpha} & 0 \\ 0 & e^{-\frac{1}{2}i\alpha} \end{bmatrix} \begin{bmatrix} \cosh \frac{1}{2}\beta & \sinh \frac{1}{2}\beta \\ \sinh \frac{1}{2}\beta & \cosh \frac{1}{2}\beta \end{bmatrix} \begin{bmatrix} e^{\frac{1}{2}i\gamma} & 0 \\ 0 & e^{-\frac{1}{2}i\gamma} \end{bmatrix}$$

This is not a representation of $SO_{2,1}$ because it changes sign when α, β, and γ are replaced by $\alpha + \pi$, $-\beta$, and $\gamma + \pi$, whereas these substitutions leave (1.32) unchanged. For the same reason the groups $SU_{1,1}$ and $SO_{2,1}$ are not isomorphic.

As already noted, the relevant representations of noncompact groups will not be finite dimensional, and so the dimension of a representation of $o_{2,1}$ cannot be used to indicate the group. In the noncompact case the eigenvalues of the compact generators provide the criterion: if $SO_{2,1}$ is generated, (1.32) shows that iD_{12} has integral eigenvalues; if $SU_{1,1}$ is generated, (1.41) with $\zeta = 0$ shows that the eigenvalues of iD_{12} are integers or half-integers.

1.12. SOME IRREDUCIBLE SKEW-HERMITIAN REPRESENTATIONS OF $o_{2,1}$

The operators (B65)

$$D_{12} \leftrightarrow -iT_3 = -\frac{iz}{2}\frac{\partial}{\partial z} + \frac{iw}{2}\frac{\partial}{\partial w}$$

$$E_{23} \leftrightarrow -iT_1 = \frac{1}{2}w\frac{\partial}{\partial z} + \frac{1}{2}z\frac{\partial}{\partial w} \qquad (1.42)$$

$$E_{31} \leftrightarrow iT_2 = -\frac{iw}{2}\frac{\partial}{\partial z} + \frac{iz}{2}\frac{\partial}{\partial w}$$

form an $o_{2,1}$ algebra, that is, satisfy (1.31). Take for their domain $\mathscr{S}$ the set of "polynomials" consisting of terms $z^a w^b$ with complex coefficients, where a is a nonnegative integer, b is a negative integer, and $b < -a$. This representation of $o_{2,1}$ will be shown below to be reducible. Each of the contained irreducible representations has a lower bound k on the eigenvalues of T_3. Only irreducible representations of this type D_k^+ (B47) are required in this book.

Shift operators for eigenvalues of T_3 are $T_+ = T_1 + iT_2 = iz\,\partial/\partial w$ and $T_- = T_1 - iT_2 = iw\,\partial/\partial z$. The function $z^a w^b$ is an eigenfunction of T_3 belonging to the eigenvalue $\frac{1}{2}a - \frac{1}{2}b$, and the action of the shift operators is

$$T_+(z^a w^b) = ib(z^{a+1}w^{b-1}), \quad T_-(z^a w^b) = ia(z^{a-1}w^{b+1}). \qquad (1.43)$$

These equations show that $\mathscr{S}$ is invariant under the operators (1.42), so that a representation of $o_{2,1}$ has been given. In particular, restricting a to be an integer prevents T_- from producing negative values of a, and the restrictions on b prevent T_- from giving nonnegative values of b (if $b = -1$, then $a = 0$). Also any subset $\mathscr{S}_k$ which has $a + b$ constant is invariant: the representation is reducible, and is the direct sum of representations defined on the $\mathscr{S}_k$. In $\mathscr{S}_k$, the Casimir operator

$$T_1{}^2 + T_2{}^2 - T_3{}^2 = T_3 - T_3{}^2 + T_+ T_-$$

has the value

$$-\tfrac{1}{4}(b - a)^2 + \tfrac{1}{2}(a - b) - (b + 1)a = -\tfrac{1}{4}(a + b)^2 - \tfrac{1}{2}(a + b) = k - k^2$$

where $k = -\tfrac{1}{2}a - \tfrac{1}{2}b$ is one of the numbers $\tfrac{1}{2}$, 1, $\tfrac{3}{2}$, 2, ..., and is uniquely determined by $k - k^2$ since $a + b < 0$. The invariant subspace is spanned by the functions $z^a w^{-a-2k}$ ($a = 0, 1, 2, \ldots$), which are eigenfunctions of T_3 belonging to the eigenvalues $k + a$. Thus k is the minimum eigenvalue of T_3. Since $b \neq 0$, the action of T_+ shows that the representation on $\mathscr{S}_k$ is irreducible. The operators are skew-Hermitian if an inner product is defined by

$$\langle z^A w^B | z^a w^b \rangle = \binom{-b - 1}{a}^{-1} \delta_{a,A}\, \delta_{b,B} \qquad (1.44)$$

which requires the previously stated restriction $b < -a$ and defines the inner product of any pair of functions in the domain because the inner product is linear in the right factor and antilinear in the left factor. Then

$$\langle z^A w^B | T_3 | z^a w^b \rangle = \tfrac{1}{2}(a - b)\binom{-b - 1}{a}^{-1} \delta_{a,A}\, \delta_{b,B}$$

and

$$\langle z^A w^B | T_3^\dagger | z^a w^b \rangle = \overline{\langle z^a w^b | T_3 | z^A w^B \rangle} = \tfrac{1}{2}(a - b)\binom{-b - 1}{a}^{-1} \delta_{a,A}\, \delta_{b,B}$$

showing that $(iT_3)^\dagger = -(iT_3)$. Similarly, writing only the nonzero cases,

$$\langle z^{a+1} w^{b-1} | T_+ | z^a w^b \rangle = ib\binom{-b}{a+1}^{-1} = -i(a + 1)\binom{-b - 1}{a}^{-1}$$

$$\langle z^{a+1} w^{b-1} | T_-^\dagger | z^a w^b \rangle = \overline{\langle z^a w^b | T_- | z^{a+1} w^{b-1} \rangle} = -i(a + 1)\binom{-b - 1}{a}^{-1}$$

showing that the shift operators are complex conjugates. This implies $(-iT_1)^\dagger = iT_1$ and $(iT_2)^\dagger = -(iT_2)$.

Other types of representation are obtained by relaxing the conditions on a and b. In particular, taking z ($a = 1, b = 0$) and w ($a = 0, b = 1$) gives the fundamental representation, for which $k = -\tfrac{1}{2}$ and the value of the Casimir operator is $-\tfrac{3}{4}$. More precisely, this representation is equivalent to the fundamental representation through the correspondence $z \leftrightarrow \begin{bmatrix} 1 \\ 0 \end{bmatrix}$, $w \leftrightarrow \begin{bmatrix} 0 \\ 1 \end{bmatrix}$.

The negative powers of w appearing in $\mathscr{S}$ can be avoided by using an equivalent representation, obtained by replacing w by w^{-1} and b by $-b$. The domain is the set of complex polynomials spanned by $z^a w^b$ where a is a non-negative integer, b is a positive integer, and $b > a$. The operators are

$$D_{12} \leftrightarrow -\frac{iz}{2}\frac{\partial}{\partial z} - \frac{iw}{2}\frac{\partial}{\partial w}, \; E_{31} + iE_{23} \leftrightarrow zw^2 \frac{\partial}{\partial w},$$

$$-E_{31} + iE_{23} \leftrightarrow w^{-1}\frac{\partial}{\partial z}$$

A subspace with $b - a$ constant is an invariant space, in which the Casimir operator has the value $k - k^2$ where $k = \frac{1}{2}b - \frac{1}{2}a$ is the minimum eigenvalue of $(z/2)(\partial/\partial z) + (w/2)(\partial/\partial w)$. If an inner product is defined by

$$\langle z^a w^b | z^A w^B \rangle \; = \; \binom{b-1}{a}^{-1} \delta_{a,A}\,\delta_{b,B}$$

the algebra is skew-Hermitian. The domain giving the fundamental representation now contains z $(a = 1, b = 0)$ and w^{-1} $(a = 0, b = -1)$.

1.13. UNITARY REPRESENTATIONS OF A SUBGROUP OF $SU_{1,1}$

The invariant spaces $\mathscr{S}_k$ of the preceding section will lead to irreducible, unitary representations of any group obtained from $o_{2,1}$ by an exponential function. For later applications it will be sufficient to obtain such representations for the subgroup $\exp(i\lambda T_2)$ generated by iT_2. Then in the fundamental representation of $o_{2,1}$ given in Section 1.11, $\varepsilon = 0, \zeta = -\bar\zeta = \frac{1}{2}\lambda i = \frac{1}{2}\theta$, and the transformations of z and w are given by the matrices

$$e^K = \begin{bmatrix} \cosh\frac{1}{2}\lambda & i\sinh\frac{1}{2}\lambda \\ -i\sinh\frac{1}{2}\lambda & \cosh\frac{1}{2}\lambda \end{bmatrix}$$

A representation of the resulting subgroup of $SU_{1,1}$ consists of the transformations

$$\exp(i\lambda T_2)(z^a w^b) = (z\cosh\tfrac{1}{2}\lambda - iw\sinh\tfrac{1}{2}\lambda)^a (iz\sinh\tfrac{1}{2}\lambda + w\cosh\tfrac{1}{2}\lambda)^b \quad (1.45)$$

defined on $\mathscr{S}$. On using the Binomial series, which is finite for a positive integer a, the transformed function becomes

$$\sum_{r=0}^{a}\binom{a}{r}(z\cosh\tfrac{1}{2}\lambda)^r(-iw\sinh\tfrac{1}{2}\lambda)^{a-r} \sum_{s=0}^{\infty}\binom{b}{s}(iz\sinh\tfrac{1}{2}\lambda)^s(w\cosh\tfrac{1}{2}\lambda)^{b-s}$$

and so contains terms $z^{r+s}w^{a+b-r-s}$ of the same degree $a + b$. Thus the subspace $\mathscr{S}_k$, invariant under the algebra (1.42), is also invariant under the transformations (1.45).

The double sum over r and s can be rearranged to $\sum_{A=0}^{\infty}\sum_r$ where $A = r + s$, and r takes the values 0 to the smaller of a and A. Using the inner product defined in the preceding section, and writing $z^a w^b = z^a w^{-a-2k} = |ba\rangle$, gives

$$\langle BA| \exp{(i\lambda T_2)}|ba\rangle = \begin{pmatrix} A + 2k - 1 \\ A \end{pmatrix}^{-1} i^{A-a} \sum_r \begin{pmatrix} a \\ r \end{pmatrix} \begin{pmatrix} -a - 2k \\ A - r \end{pmatrix}$$

$$\times (\cosh{\tfrac{1}{2}\lambda})^{-a-A-2k+2r} (\sinh{\tfrac{1}{2}\lambda})^{A+a-2r}$$

in which the first factor appears because $|ba\rangle$ is not normalized. Since

$$\begin{pmatrix} a \\ r \end{pmatrix} = \frac{a(a-1)\cdots(a-r+1)}{r!}$$

and

$$\begin{pmatrix} -a - 2k \\ A - r \end{pmatrix} = \frac{(-a-2k)(-a-2k-1)\cdots(-a-A-2k+r+1)}{(A-r)!}$$

the summation contains

$$(-)^r(-a)(-a+1)\cdots(-a+r-1)(-)^r(-A)(-A+1)\cdots(-A+r-1)/r!\,A!$$

while the numerator of $\begin{pmatrix} -a - 2k \\ A - r \end{pmatrix}$ can be written as

$$\frac{(a+2k)(a+2k+1)\cdots(a+2k+A-1)(-)^A}{(-a-A-2k+1)(-a-A-2k+2)\cdots(-a-A-2k+r)}$$

These manipulations show that

$$\langle BA| \exp{(i\lambda T_2)}|ba\rangle$$

$$= \begin{pmatrix} A + 2k - 1 \\ A \end{pmatrix}^{-1} \frac{(-i\sinh{\tfrac{1}{2}\lambda})^{a+A}(a+A+2k-1)!}{(\cosh{\tfrac{1}{2}\lambda})^{a+A+2k} A!\,(a+2k-1)!} F \qquad (1.46)$$

where F is a hypergeometric function:

$$F = {}_2F_1(-a, -A; -a-A-2k+1; \coth^2{\tfrac{1}{2}\lambda})$$

Although $-a - A - 2k + 1$ is a negative integer, it is less than $-a$ or $-A$, so the hypergeometric series terminates before a zero appears in a denominator. Putting

$$\begin{pmatrix} A + 2k - 1 \\ A \end{pmatrix}^{-1} = \frac{A!\,(2k-1)!}{(A+2k-1)!}$$

gives an expression symmetric in a and A:

$$\langle ba| \exp{(i\lambda T_2)}|BA\rangle = \langle BA| \exp{(i\lambda T_2)}|ba\rangle.$$

Changing the sign of λ or taking the complex conjugate (λ real) multiplies the expression by $(-)^{a+A}$. These two results together verify that $\exp{(i\lambda T_2)}$ is unitary.

1.14. GROUP THEORY AND QUANTUM MECHANICS

The first applications of group theory to quantum mechanics utilized obvious geometrical symmetries of physical systems. In a central field problem, the potential energy depends only on the distance from the center, and so the potential function is independent of the way in which rectangular coordinate axes are chosen. The energy operator then commutes with all the transformation operators R of the three-dimensional rotation group SO_3. If ψ is an energy eigenfunction, $R\psi$ is also an energy eigenfunction, belonging to the same eigenvalue, so that the set of eigenfunctions belonging to a fixed energy eigenvalue is invariant under SO_3. This function space can therefore be taken as the domain $\mathscr{S}$ of a representation $[\psi]$ of this group. With respect to the usual inner product ($\|\psi\|^2 = \int |\psi|^2\, dV$) the representation is unitary and is either irreducible or a direct sum of irreducible representations. Those physical properties of the system which are consequences of the rotational symmetry can be obtained from the representation theory of SO_3. The latter is independent of the particular central potential under consideration, and can therefore be worked out once, and then applied in a variety of different physical cases.

In practice this program can be achieved almost completely using properties of the algebra o_3, since the domain $\mathscr{S}$ of degenerate energy eigenfunctions also leads to a representation of the algebra. The results—the quantum theory of angular momentum—are reviewed in Chapter 2. This work is also relevant if the symmetry transformations form the group SU_2. If $[\psi]$ is irreducible, all the eigenfunctions belonging to one energy can be obtained from one such eigenfunction by applying rotation operators. The degeneracy of the energy eigenvalue is then explained completely by the rotational symmetry. If the representation of SO_3 is reducible, as in the Coulomb problem, the degeneracy is not entirely due to rotational symmetry. At first this situation was merely described as accidental degeneracy and not investigated, but in 1935 Fock discovered a larger set of operators commuting with the energy, this set forming the group SO_4. The degeneracy is explained completely by this higher symmetry: the energy eigenfunctions belonging to one eigenvalue form the domain of an irreducible representation of SO_4 or o_4. This suggests that any accidental degeneracy is caused by a hidden symmetry. In general the group for which the degenerate states belonging to single energies give irreducible representations is called the symmetry group or invariance group of the system, and the corresponding algebra is the invariance algebra.

The energy operator is a Casimir operator of the invariance algebra; if C_k are standard Casimir operators, such as $\sum g_{ij} E_i E_j$, the energy must be some function of these C_k. The Casimir operators of subalgebras can also be chosen to be Hermitian and physically significant. Quantum-mechanical calculations often require a basis to be chosen in $\mathcal{S}$ by taking the simultaneous eigenfunctions of a complete set Γ of commuting observables. If one of the Γ is a Casimir operator C_i of a subalgebra $\mathcal{L}_i$, then the eigenfunctions of C_i belonging to a fixed eigenvalue are invariant under $\mathcal{L}_i$. The resulting basis divides $\mathcal{S}$ into subspaces invariant under $\mathcal{L}_i$, so that the reducible representation of $\mathcal{L}_t$ gets decomposed into a direct sum. Any state belonging to (the domain of) an irreducible representation will usually be orthogonal to all states of (the domain of) an inequivalent representation, since the Casimir operator will have different eigenvalues in the two representations.

A basis chosen according to irreducible representations of subalgebras of the invariance algebra can be regarded as a group-theoretical classification of degenerate energy eigenstates. One advantage of such a basis is that the matrix elements of any operator belonging to the invariance algebra or group, or of any Casimir operator, are then known from representation theory, provided this theory includes a suitably defined inner product. Another way of saying this is that these matrix elements can be calculated using any representation which is isometrically equivalent to $[\psi]$. For example, the matrix elements of rotation operators relative to the basis of spherical harmonics, the $D^l_{mn}(\alpha, \beta, \gamma)$ given at the end of Section 1.7, are usually calculated from the equivalent representation on the space of homogeneous polynomials of degree l in two complex variables.

These methods can be extended by discovering larger algebras and groups for which some sets of quantum-mechanical operators and states give an irreducible representation. If the domain contains eigenfunctions of different energies, the algebra will be called a noninvariance algebra of the system. The representation theory of such an algebra may be useful for calculating nonzero matrix elements between states of different energy. For this purpose each operator of physical interest should belong to the algebra or be a function of a single element of the algebra. This consideration determines the algebras designated later as the noninvariance algebras of various systems. The domain consisting of all bound state wave functions provides a reducible representation.

Operators for the noninvariance algebra may be suggested by a knowledge of the representations of the various possible algebras. The spherical harmonics provide an example. How should one define operators which shift l and do not change m? Suppose $m \geq 0$. The basis functions Y_l^m ($l = m, m + 1, \ldots$) are specified by one quantum number l, so that an algebra of rank 1 should be sufficient. An operator with eigenvalues a function of l, and the two shift operators, will form an algebra of dimension at least 3, suggesting $c*o_3$. The subset of Hermitian operators will then form either an

o_3 algebra or an $o_{2,1}$ algebra. However, the irreducible Hermitian representations of o_3 are all finite dimensional, so if the shift operators are going to connect all the spherical harmonics of a given m, then an $o_{2,1}$ algebra is indicated. Let $S_1 \leftrightarrow iE_{23}$, $S_2 \leftrightarrow -iE_{31}$, and $S_3 \leftrightarrow iD_{12}$. Then $S_\pm = S_1 \pm iS_2$ shift the eigenvalues of S_3 by ± 1, and so these eigenvalues should be $l + q$ with q real; as they are bounded below by $m + q$, the representation of $o_{2,1}$ will be of the type D_k^+ considered in Section 1.12. Putting $q = k - m$, the minimum eigenvalue of S_3 is k, S_3 is defined by $S_3 Y_l^m = (l - m + k)Y_l^m$, and $S_3 \leftrightarrow T_3$ with $a = l - m$. Definitions of $S_\pm$ now follow from the correspondence $S_\pm \leftrightarrow T_\pm$ with the representation in Section 1.12, but (1.43) does not apply immediately because $z^a w^b$ is not normalized. From (1.44), the isometric correspondence between domains is

$$Y_l^m \leftrightarrow \left(\frac{2k + l - m - 1}{l - m}\right)^{1/2} z^{l - m} w^{m - l - 2k}$$

and so (1.43) gives

$$\begin{aligned} S_+ Y_l^m &= -i(l - m + 1)^{1/2}(2k + l - m)^{1/2} Y_{l+1}^m \\ S_- Y_l^m &= i(l - m)^{1/2}(2k + l - m - 1)^{1/2} Y_{l-1}^m \end{aligned} \tag{1.47}$$

Provided $k \neq 0$, $-\frac{1}{2}$, -1, $-\frac{3}{2}$, ..., these definitions of S_3, $S_1 = \frac{1}{2}S_+ + \frac{1}{2}S_-$, and $S_2 = \frac{1}{2}iS_- - \frac{1}{2}iS_+$ give a (Hermitian) representation of $o_{2,1}$ of type D_k^+. Taking $k = m + \frac{1}{2}$ gives $S_3 = S$, $S_+ = -iA_0^\dagger$, $S_- = iA_0$, operators of the $o_{3,2}$ algebra of Section 1.10. The $o_{2,1}$ algebra is the subalgebra of transformations of x_3, x_4, and x_5 ($D_{45} \leftrightarrow -iS_3$, $E_{35} \leftrightarrow iS_1$, $E_{34} \leftrightarrow -iS_2$).

1.15. REPRESENTATIONS ON AN ALGEBRA

In the representation of o_3 by vector algebra, mentioned in Section 1.6, vectors both represent the operators and form the domain. The cross product gives both the Lie product and the rule defining the action of the representative of an operator. Any $\mathscr{L}$ algebra of operators gives a representation of $\mathscr{L}$ in the same way: the operators also form the domain $\mathscr{S}$ of the representation, and A, an operator representing an element of $\mathscr{L}$, changes F into $[A, F]$, both F and $[A, F]$ belonging to the domain. Here it is preferable perhaps to use the term representation space for the domain of the representation. The concept of invariant subspace, hence of irreducible or reducible representation, still applies. As the domain should be a complex linear space, $\mathscr{S}$ will actually be the complexification of the algebra. Other domains are also possible. Quantum-mechanical operators forming a representation space of an $\mathscr{L}$ algebra are called a tensor operator of the $\mathscr{L}$ algebra, each operator being a component of the tensor. The tensor operator is irreducible if no subset of the components is itself invariant under the $\mathscr{L}$ algebra.

Similarly a set of operators can form a representation space for a group: A changes F into AFA^{-1}.

Examples of representations on an algebra can be seen in the tables of Appendix D. The first three rows of Table 1 show the $c*o_{3,2}$ algebra as a representation space for o_3. This ten-dimensional representation is the direct sum of four irreducible representations. One of these is one-dimensional: the representation space is all complex multiples of S, including zero, and any element of this space is changed to 0 by any operator of o_3. The other three irreducible representations are equivalent: their representation spaces are spanned by the L_i, the U_i, and the V_i. Similarly the first six rows and last four columns of Table 1 show that S and the V_i span a representation space for $o_{3,1}$. This four-dimensional representation is irreducible and equivalent to the representation on the domain of linear functions through the correspondence $x_1 \leftrightarrow V_1, x_2 \leftrightarrow V_2, x_3 \leftrightarrow V_3, x_4 \leftrightarrow -S$. The contained representation of the o_3 subalgebra is a direct sum of two irreducible representations.

Evidently an algebra is always a representation space for any subalgebra, and the representation is always reducible since the subalgebra itself is an invariant subspace of the representation space.

Examples appearing in Table 2 of Appendix D include the following. The first ten rows and last five columns show the W_i and T as the basis of a representation space of $o_{4,1}$, giving a representation equivalent to that on linear functions. Representation spaces of o_4 of dimensions 6, 4, 4, and 1 are apparent from the first six rows. The first of these is reducible, since the components of $\mathbf{K} = \frac{1}{2}\mathbf{L} + \frac{1}{2}\mathbf{M}$ and $\mathbf{N} = \frac{1}{2}\mathbf{L} - \frac{1}{2}\mathbf{M}$ are invariant. Any four-dimensional, irreducible representation of o_4 is equivalent to that on linear functions; for example $x_i \leftrightarrow V_i$ or W_i. An irreducible four-dimensional tensor of o_4 is called a 4-vector, and the components corresponding to the x_i are its Cartesian components.

1.16. PRODUCT REPRESENTATIONS

Suppose that $[F]$ and $[H]$ are representations of a group $\mathscr{G}$, with ρ_F and ρ_H representing ρ of $\mathscr{G}$, and suppose that a product is defined for elements f and h of the domains $\mathscr{F}$ and $\mathscr{H}$ of the two representations. Then the domain $\mathscr{S}$ of the product representation $[F] \times [H]$ is the linear space containing all the products fh, and ρ is represented by the operator changing fh to $(\rho_F f)(\rho_H h)$. An important problem of representation theory is to express $[F] \times [H]$ as a direct sum of irreducible representations. This problem can also be formulated in terms of the product group $\mathscr{G} \times \mathscr{G}$, which consists of the set of pairs (ρ, τ) with the multiplication rule $(\rho, \tau)(\omega, \chi) = (\rho\omega, \tau\chi)$, where ρ, τ, ω, and χ are all elements of $\mathscr{G}$. The subgroup of all (ρ, ρ) is isomorphic to $\mathscr{G}$. In the representation of $\mathscr{G} \times \mathscr{G}$ on $\mathscr{S}$ in which (ρ, τ) is represented by the operator changing fh to $(\rho_F f)(\tau_H h)$, the subgroup has the representation $[F] \times [H]$. The problem is then that of decomposing $\mathscr{S}$ into the subspaces which are invariant under the subgroup.

The corresponding concepts for Lie algebras are slightly different because multiplication in the group corresponds to addition in the algebra. Thus if A_F and A_H represent A of an algebra $\mathscr{L}$, then in the product representation the representative of A is the operator changing fh to $(A_F f)h + f(A_H h)$. The algebra $\mathscr{L} \oplus \mathscr{L}$ consisting of pairs (A, B) with the definitions $(A, B) + (C, D) = (A + C, B + D)$, $\lambda(A, B) = (\lambda A, \lambda B)$, and $[(A, B), (C, D)] = ([A, C], [B, D])$ is called the direct sum, and its subalgebra consisting of all (A, A) is isomorphic to $\mathscr{L}$. When $\mathscr{L} \oplus \mathscr{L}$ is represented on $\mathscr{S}$ with the representative of (A, B) changing fh to $(A_F f)h + f(B_H h)$, the subalgebra has the representation $[F] \times [H]$.

As a first example, suppose $[F]$ is the fundamental 2×2 matrix representation of $o_{2,1}$, and $[H]$ is the representation on linear functions. Then $\mathscr{F}$ is $\mathscr{V}_2$, spanned by $\alpha = \begin{bmatrix} 1 \\ 0 \end{bmatrix}$ and $\beta = \begin{bmatrix} 0 \\ 1 \end{bmatrix}$, and $\mathscr{H}$ is spanned by the functions x, y, and z. These representations of $o_{2,1}$ are summarized by the table below.

	α	β	x	y	z
D_{12}	$-\frac{1}{2}i\alpha$	$\frac{1}{2}i\beta$	y	$-x$	0
E_{23}	$\frac{1}{2}\beta$	$\frac{1}{2}\alpha$	0	z	y
E_{31}	$-\frac{1}{2}i\beta$	$\frac{1}{2}i\alpha$	z	0	x

Thus, for example, the representative of E_{23} in $[F]$ changes β to $\frac{1}{2}\alpha$. The domain $\mathscr{S}$ of the product representation is spanned by αx, βx, αy, βy, αz, and βz with $\alpha x = \begin{bmatrix} x \\ 0 \end{bmatrix}$, $\beta x = \begin{bmatrix} 0 \\ x \end{bmatrix}$, etc., and the action of the representatives of D_{12}, E_{23}, and E_{31} is:

	αx	αy	αz
D_{12}	$-\frac{1}{2}i\alpha x + \alpha y$	$-\frac{1}{2}i\alpha y - \alpha x$	$-\frac{1}{2}i\alpha z$
E_{23}	$\frac{1}{2}\beta x$	$\frac{1}{2}\beta y + \alpha z$	$\frac{1}{2}\beta z + \alpha y$
E_{31}	$-\frac{1}{2}i\beta x + \alpha z$	$-\frac{1}{2}i\beta y$	$-\frac{1}{2}i\beta z + \alpha x$

	βx	βy	βz
D_{12}	$\frac{1}{2}i\beta x + \beta y$	$\frac{1}{2}i\beta y - \beta x$	$\frac{1}{2}i\beta z$
E_{23}	$\frac{1}{2}\alpha x$	$\frac{1}{2}\alpha y + \beta z$	$\frac{1}{2}\alpha z + \beta y$
E_{31}	$\frac{1}{2}i\alpha x + \beta z$	$\frac{1}{2}i\alpha y$	$\frac{1}{2}i\alpha z + \beta x$

The functions $\alpha z + i\beta x - \beta y$ and $\alpha x - i\alpha y + i\beta z$ span an invariant subspace; the product representation is reducible into two irreducible representations of dimensions 2 and 4.

A second example uses Table 1 of Appendix D. The operators $-iS \leftrightarrow D_{12}$, $-U_3 \leftrightarrow E_{23}$, and $V_3 \leftrightarrow E_{31}$ form an $o_{2,1}$ algebra. Suppose that $\mathscr{F}$ is spanned by L_1, U_2, and V_2, and $\mathscr{H}$ is spanned by L_2, U_1, and V_1. Table 1 shows that these representation spaces are invariant under the commutation operation and so define representations of $o_{2,1}$. Then $\mathscr{S}$ is spanned by the nine products $L_1 L_2$, $L_1 U_1$, $L_1 V_1$, $U_2 L_2$, $U_2 U_1$, $U_2 V_1$, $V_2 L_2$, $V_2 U_1$, and $V_2 V_1$. Since, for example, $[V_3, L_1]L_2 + L_1[V_3, L_2]$ is $[V_3, L_1 L_2]$, the action of the representatives is still to take the commutator. In general if both $[F]$ and $[H]$ are defined on an algebra, and $A_F = A_H$, this is also the representative of A in $[F] \times [H]$.

If $[F]$ is defined on an algebra and $[H]$ on a function space, the product fh is a function evaluated by allowing the operator f to act on the function h. Suppose that the representatives A_F and A_H are the same or differ only by having different domains, so that $A_F h = A_H h$. Then the operation on fh gives $(A_F f - f A_F)h + f(A_H h) = A_F(fh)$. The representative of A in $[F] \times [H]$ is also A_F. (More precisely, as the domains are different, the representatives in $[F]$, $[H]$ and $[F] \times [H]$ have the same rule.)

In the third example, then, representatives of $o_{2,1}$ will be given by (1.42), $[F]$ is defined by the representation space $\mathscr{F}$ spanned by $\partial/\partial z$ and $\partial/\partial w$, and $[H]$ is D_0^+ of Section 1.12 so that $\mathscr{H}$ is the space spanned by all polynomials in (z/w). The domain $\mathscr{S}$ of the product representation contains all the products

$$\frac{\partial}{\partial z}\left(\frac{z}{w}\right)^a \quad \text{and} \quad \frac{\partial}{\partial w}\left(\frac{z}{w}\right)^a$$

These are not all independent, because

$$\frac{\partial}{\partial w}\left(\frac{z}{w}\right)^{a-1} = \frac{1-a}{a}\frac{\partial}{\partial z}\left(\frac{z}{w}\right)^a$$

Thus $\mathscr{S}$ is spanned by the functions $z^a w^{-a-1}$ $(a = 0, 1, 2, 3, \ldots)$, and $[F] \times [H]$ is $D_{1/2}^+$ of Section 1.12.

The spherical harmonic addition theorem can be interpreted in terms of a product representation. Products $Y_l^m(\theta_1, \phi_1)Y_l^k(\theta_2, \phi_2)$ are the domain $\mathscr{S}$ of the l^2-dimensional representation $[l] \times [l]$ of SO_3. This representation is reducible, and (1.28) gives the invariant subspace of dimension 1. Here the operators ρ_F, transformations of θ_1 and ϕ_1, commute with the operators τ_H, which are transformations of θ_2 and ϕ_2. It is then permissible to use $\mathscr{S}$ as the domain of the ρ_F and the τ_H, and represent the product group by the operators $\rho_F \tau_H$.

Similarly, for an algebra, $\mathscr{S}$ can be used as the domain of A_F and B_H

provided that these operators commute, and the element (A, B) of $\mathscr{L} \oplus \mathscr{L}$ is then represented by $A_F + B_H$. For instance, it was observed in Section 1.2 that o_4 is the sum of two o_3 algebras. A representation $[\Xi]$ of o_3 is given by the components of $-i\mathbf{K}$ on the domain of hyperspherical harmonics. Another representation $[T]$ is given by the components of $-i\mathbf{N}$. These operators commute, and so the sums $-i(K_\alpha + N_\beta)$ give a representation of $o_3 \oplus o_3 = o_4$. The subalgebra consisting of sums of corresponding components, that is, $\alpha = \beta$ giving $-i(\mathbf{K} + \mathbf{N})$, gives a representation $[\Xi] \times [T]$ of o_3, which is decomposed into irreducible representations by using the Y_{nlm} as a basis of the domain.

2. Angular Momentum Theory

Chapter 3 will show that the symmetry group of the Coulomb problem is generated by the algebra o_4 which is $o_3 \oplus o_3$. Also calculations requiring noninvariance groups usually depend on subgroups generated by $o_{2,1}$ algebras, which have the same complexification as o_3. Consequently nearly all algebraic calculations on the Coulomb problem can be carried through in terms of angular momentum theory. This mode of presentation is adopted in this book, on the assumption that readers have met the quantum theory of angular momentum.

The main purpose of Chapter 2 is to summarize those parts of angular momentum theory to be used and to give the modifications required to treat representations of $o_{2,1}$ of type D_k^+. No knowledge of groups or algebras is necessary to read this theory. Its relation to the work in Chapter 1 is described in Section 2.4.

2.1. EIGENVALUES AND EIGENSTATES

In quantum mechanics the Hermitian operators J_x, J_y, and J_z are said to be the components of an angular momentum $\mathbf{J}$ if they have the commutation relations

$$[J_x, J_y] = iJ_z, \quad [J_y, J_z] = iJ_x, \quad [J_z, J_x] = iJ_y \tag{2.1}$$

This definition generalizes the example provided by (1.9), for which (1.7) shows that hL_x, hL_y, and hL_z are the operators representing the components of the angular momentum $\mathbf{r} \times \mathbf{p}$.

From (2.1), any component of the angular momentum commutes with $J^2 = J_x^2 + J_y^2 + J_z^2$, and so there are simultaneous eigenstates of J^2 and one component, for which J_z is usually chosen. Because the components are Hermitian operators defined on a positive-definite space, the possible eigenvalues of J^2 are $j(j+1)$, where j is a nonnegative integer or half-integer. The possible eigenvalues of any component are j. Eigenvalues of J_z are usually denoted by m. If a simultaneous eigenstate of J^2 and J_z belongs to the respective eigenvalues $j(j+1)$ and m, then

$$|m| \leqslant j, \quad \text{and} \quad j - m \text{ is an integer} \tag{2.2}$$

The simultaneous eigenstate is written $|\gamma jm\rangle$ where γ denotes any other quantum numbers needed to specify the state, for example, the principal quantum number n of Coulomb energy eigenfunctions. The γ will usually be omitted in the formulas of this chapter.

The operators $J_\pm = J_x \pm iJ_y$ are called raising and lowering operators because $J_\pm |jm\rangle$ are eigenstates of J_z belonging to the eigenvalues $m \pm 1$ (unless $m = \pm j: J_\pm |j \pm j\rangle = 0$). So for given j, there are eigenstates belonging to each of the $2j + 1$ values of $m(m = -j, -j + 1, \ldots, j - 1, j)$ allowed by (2.2). The degeneracy of the eigenvalue $j(j + 1)$ is therefore $(2j + 1)N$ where the integer N is the number of associated values of γ. If the $|j\ m\rangle$ are normalized, their relative phase can be chosen so that

$$J_\pm |j\ m\rangle = (j \mp m)^{1/2}(j \pm m + 1)^{1/2}|j\ m \pm 1\rangle \qquad (2.3)$$

The operators (1.9) are an example of an orbital angular momentum, for which only integer values of j and m occur. Then l is usually used instead of j. For the case (1.9), the spherical harmonics (1.13) are a realization of the eigenstates $|l\ m\rangle$, and (1.35) is (2.3).

2.2.   ADDITION OF ANGULAR MOMENTA

The $\mathbf{J}_1$ and $\mathbf{J}_2$ are commuting angular momenta if each component of $\mathbf{J}_1$ commutes with every component of $\mathbf{J}_2$. Then there are simultaneous eigenstates of $J_1{}^2, J_{1z}, J_2{}^2$, and J_{2z} ; these eigenstates are denoted by $|j_1\ m_1\ j_2\ m_2\rangle$ and will satisfy (2.3) for both $J_{1\pm}$ and $J_{2\pm}$. Often $\mathbf{J}_1$ and $\mathbf{J}_2$ commute because they involve independent coordinates, and the wave function for $|j_1\ m_1\ j_2\ m_2\rangle$ is then just the product of wave functions corresponding to $|j_1\ m_1\rangle$ and $|j_2\ m_2\rangle$. Thus if $(r_1\ \theta_1\ \phi_1)$ and $(r_2\ \theta_2\ \phi_2)$ are spherical polar coordinates of two particles, and their orbital angular momenta are $\mathbf{L}_i = \mathbf{r}_i \times \mathbf{p}_i$, then the functions $Y_{l_1}^{m_1}(\theta_1, \phi_1)\ Y_{l_2}^{m_2}(\theta_2, \phi_2)$ are simultaneous eigenfunctions of $L_1{}^2, L_{1z}, L_2{}^2$, and L_{2z}. However in the application to the Coulomb problem the situation is different, because the commuting angular momenta $\mathbf{K}$ and $\mathbf{N}$ appearing in (1.11) satisfy $K^2 = N^2$. The expressions for $\mathbf{K}$ and $\mathbf{N}$ in Appendix C contain the same coordinates, and the eigenfunctions (1.17) are not products in the sense used here. Because $K^2 = N^2$ the results given below will only be needed for the case $j_1 = j_2$.

The commutation relations of $J_x = J_{1x} + J_{2x}$, etc., show that $\mathbf{J} = \mathbf{J}_1 + \mathbf{J}_2$ is also an angular momentum, and so there are also simultaneous eigenstates of J^2 and J_z. These two operators commute with $J_1{}^2$ and $J_2{}^2$, but J^2 does not commute with J_{1z} and J_{2z}. Thus there are simultaneous eigenstates $|j_1\ j_2\ J\ M\rangle$ of $J_1{}^2, J_2{}^2, J^2$, and J_z, but these are not eigenstates of J_{1z} and J_{2z}. For given j_1 and j_2, the values of J are $j_1 + j_2, j_1 + j_2 - 1, \ldots, |j_1 - j_2|$, and for each value of J there are the usual $2J + 1$ values of M. This means $(2j_1 + 1)(2j_2 + 1)$ values of JM, and these orthonormal states $|j_1 \, j_2 \, J \, M\rangle$ are linear combinations of the $(2j_1 + 1)(2j_2 + 1)$ orthonormal states $|j_1 \, m_1 \, j_2 \, m_2\rangle$, the transformation being unitary. The transformation coefficients $\langle j_1 \, m_1 \, j_2 \, m_2 | j_1 \, j_2 \, JM\rangle = \overline{\langle j_1 \, j_2 \, J \, M | j_1 \, m_1 \, j_2 \, m_2\rangle}$ are called Clebsch-Gordan coefficients. The phases of the states $|j_1 \, j_2 \, J \, M\rangle$ are conventionally chosen so that the coefficients are all real, so the bar can be omitted, and the unitary transformation is orthogonal. Also, because $J_z = J_{1z} + J_{2z}$, the coefficients are zero unless $M = m_1 + m_2$, and so the $(2j_1 + 1)(2j_2 + 1)$-dimensional transformation matrix is the direct sum of $2j_1 + 2j_2 + 1$ matrices corresponding to the different values of M. The transformation may therefore be written

$$|j_1 \, j_2 \, J \, M\rangle = \sum_{m_1} |j_1 \, m_1 \, j_2 \, m_2 = M - m_1\rangle\langle j_1 \, m_1 \, j_2 \, m_2 | j_1 \, j_2 \, J \, M\rangle \quad (2.4)$$

$$|j_1 \, m_1 \, j_2 \, m_2\rangle = \sum_{J} |j_1 \, j_2 \, J \, M = m_1 + m_2\rangle\langle j_1 \, m_1 \, j_2 \, m_2 | j_1 \, j_2 \, J \, M\rangle \quad (2.5)$$

where the sum over J is from $J = |M|$ to $J = j_1 + j_2$.

The conventional phase choice for the Clebsch-Gordan coefficients is to make

$$\langle j_1 \, m_1 = j_1 \, j_2 \, m_2 = J - j_1 | j_1 \, j_2 \, J \, M = J\rangle > 0 \quad (2.6)$$

This convention, together with (2.3), determines unique real values for all the coefficients. This is obvious from a method of calculating the coefficients, which will be illustrated for the case $j_1 = j_2 = 1$. There is only one state with the maximum value of M, so $|J = M = 2\rangle = P|m_1 = m_2 = 1\rangle$ where P is a phase factor. So $|P| = 1$, and (2.6) gives $P = 1$. Using (2.3) to evaluate each side of $J_- |J = M = 2\rangle = J_{1-}|m_1 = m_2 = 1\rangle + J_{2-}|m_1 = m_2 = 1\rangle$ gives

$$\sqrt{2}|J = 2 \, M = 1\rangle = |m_1 = 1 \, m_2 = 0\rangle + |m_1 = 0 \, m_2 = 1\rangle \quad (2.7)$$

Three further applications of the lowering operators $J_- = J_{1-} + J_{2-}$ give the states $|J = 2 \, M\rangle$ for $M = 0, -1$, and -2, and all the coefficients for the case $j_1 = j_2 = 1, J = 2$ have been calculated. They must be real and positive, since the numerical factor in (2.3) is positive. There are two states with $M = 1$ ($J = 2, 1$); since $|J = M = 1\rangle$ is normalized, and orthogonal to the known state $|J = 2 \, M = 1\rangle$, it can be obtained from (2.7) up to a phase factor P:

$$\sqrt{2}|J = M = 1\rangle = P(|m_1 = 1 \, m_2 = 0\rangle - |m_1 = 0 \, m_2 = 1\rangle)$$

Equation (2.6) now gives $P = 1$. Two applications of J_- using (2.3) now give all the remaining coefficients for the case $j_1 = j_2 = J = 1$. Finally the remaining state $|J = M = 0\rangle$ is determined except for phase by normalization and by orthogonality to the known states $|J = 2 \, M = 0\rangle$ and $|J = 1 \, M = 0\rangle$. The phase, determined by (2.6), requires the relevant coefficients to be real. The method applies for any values of j_1 and j_2.

The order of addition of angular momenta has no physical significance, and so the states $|j_1 \ j_2 \ J \ M\rangle$ and $|j_2 \ j_1 \ J \ M\rangle$ can only differ by a phase factor, which must be ± 1 since the Clebsch-Gordan coefficients are real. The states need not be identical, because the phase convention (2.6) distinguishes the first angular momentum quantum numbers $j_1 \ m_1$ from the second angular momentum quantum numbers $j_2 \ m_2$. They are identical only when $j_1 + j_2 - J$ is even, and have opposite sign when $j_1 + j_2 - J$ is odd:

$$\langle j_2 \ m_2 \ j_1 \ m_1 | j_2 \ j_1 \ J \ M\rangle = (-)^{j_1 + j_2 - J}\langle j_1 \ m_1 \ j_2 \ m_2 | j_1 \ j_2 \ J \ M\rangle \qquad (2.8)$$

The explicit expression (W31–59) for the general Clebsch-Gordan coefficient contains a single summation which can be written as a hypergeometric function of unit argument:

$$\langle j_1 \ m_1 \ j_2 \ m_2 | j_1 \ j_2 \ J \ M\rangle$$

$$= \left[\frac{(J + j_2 - j_1)!(J + M)!(j_1 + m_1)!(j_2 + m_2)!(2J + 1)}{\begin{array}{c}(J - j_2 + j_1)!(j_1 + j_2 - J)!(j_1 + j_2 + J + 1)! \\ \times (J - M)!(j_1 - m_1)!(j_2 - m_2)!\end{array}} \right]^{1/2}$$

$$\times \, \delta_{M, m_1 + m_2}(-)^{j_1 - m_1} \frac{(j_1 + j_2 - M)!}{(j_2 - j_1 + M)!}$$

$$\times \, {}_3F_2\left[\begin{array}{ccc} J + M + 1 & -J + M & -j_1 + m_1 \\ j_2 - j_1 + M + 1 & & -j_1 - j_2 + M \end{array} ; 1\right] \qquad (2.9)$$

The hypergeometric function is defined even when the lower two parameters are negative, because the inequalities $-j_1 - j_2 + M \le -J + M$, $j_2 - j_1 + M + 1 \le J + M + 1$ ensure that the series terminate before the appearance of a zero in any denominator. The equation (2.9) differs from the original result of this type, given by Rose (R55), by a transformation of the ${}_3F_2$.

2.3. SCALAR AND VECTOR OPERATORS

An operator S is scalar with respect to an angular momentum $\mathbf{J}$ if it commutes with all components of $\mathbf{J}$. Then $S|\gamma J M\rangle$ is either zero, or is another eigenstate of J^2 and J_z belonging to the same eigenvalues, and so

$$\langle \gamma J M | S | \gamma' J' M'\rangle$$

is zero unless $J = J'$ and $M = M'$. This matrix element is also independent of M.

The operators V_x, V_y, V_z are the Cartesian components of a vector operator **V** with respect to the angular momentum **J** if the following commutation relations hold:

$$[J_x, V_y] = iV_z, [J_y, V_z] = iV_x, [J_z, V_x] = iV_y,$$
$$[J_x, V_x] = [J_y, V_y] = [J_z, V_z] = 0 \qquad (2.10)$$

Then the identity (1.4) gives also $[V_z, J_x] = iV_y$, $[V_x, J_y] = iV_z$, and $[V_y, J_z] = iV_x$.

For example, **r**, **p**, and $\mathbf{L} = \mathbf{r} \times \mathbf{p}$ are all vector operators with respect to **L**. More generally V, V', and V'' are components of a vector operator if their commutators with the components of **J** are linear combinations of V, V', and V''. Then some linear combinations of V, V', and V'' will be Cartesian components, but it is more useful to use the spherical components defined by

$$V_{\pm 1} = \mp(V_x \pm iV_y)\sqrt{\tfrac{1}{2}}, V_0 = V_z \qquad (2.11)$$

The spherical components r_m of **r** are proportional to the spherical harmonics with $l = 1 : \tfrac{1}{2}\sqrt{3} \ r_m = \pi^{1/2} r Y_1{}^m$, where $r = (x^2 + y^2 + z^2)^{1/2}$.

If V_m is a spherical component of a vector operator with respect to **J**, then $V_m|\gamma JM\rangle$ is either zero or another eigenstate of J_z belonging to the eigenvalue $M + m$. So $\langle \gamma JM|V_m|\gamma'J'M'\rangle$ is zero unless $M = m + M'$. The Wigner-Eckart theorem states that the dependence of this matrix element on m and M is the same for any vector operator **V** and has the form of a Clebsch-Gordan coefficient:

$$\langle \gamma \ J \ M|V_m|\gamma' \ J' \ M-m\rangle =$$
$$(2J + 1)^{-1/2}\langle\gamma J||\mathbf{V}||\gamma'J'\rangle\langle J' \ M-m \ 1 \ m|J' \ 1 \ J \ M\rangle \qquad (2.12)$$

This implies that nonzero matrix elements are restricted to those values of J' and J which can appear in the Clebsch-Gordan coefficient, that is, $J' = J$, $J \pm 1$. The $\langle\gamma J||\mathbf{V}||\gamma'J'\rangle$ is called the reduced (or double-bar) matrix element of **V**, and the factor $(2J + 1)^{-1/2}$ is conventional. One example of a vector operator is **J** itself, for which

$$\langle\gamma J||\mathbf{J}||\gamma'J'\rangle = \delta_{\gamma,\gamma'} \ \delta_{J, J'} \ [J(J + 1)(2J + 1)]^{1/2} \qquad (2.13)$$

Substituting this and the results of Section 2.2 into (2.12) gives the matrix representation (1.21) of $-iJ_z$, $-iJ_x = i(J_1 - J_{-1})\sqrt{\tfrac{1}{2}}$, $-iJ_y = (J_1 + J_{-1})\sqrt{\tfrac{1}{2}}$. For the example obtained by taking the matrix elements of components of **r** with respect to eigenfunctions of $\mathbf{L} = \mathbf{r} \times \mathbf{p}$, the Clebsch-Gordan coefficient appears from the integration over polar angles:

$$\int_0^\pi \sin\theta \ d\theta \int_0^{2\pi} d\phi \, \overline{Y}_l{}^M(\theta, \phi) \frac{r_m}{r} Y_{l+1}^{M-m}(\theta, \phi)$$
$$= -(l + 1)^{1/2}(2l + 1)^{-1/2}\langle l + 1 \ M - m \ 1 \ m|l + 1 \ 1 \ l \ M\rangle \qquad (2.14)$$

Let $\mathbf{J}_1$ and $\mathbf{J}_2$ be commuting angular momenta with sum $\mathbf{J}$. If $\mathbf{V}$ is a vector with respect to $\mathbf{J}_2$, and commutes with $\mathbf{J}_1$ (i.e., all components of $\mathbf{V}$ and $\mathbf{J}_1$ commute), then $\mathbf{V}$ is a vector with respect to $\mathbf{J}$. The reduced matrix element of $\mathbf{V}$ with respect to the states $|j_1\ j_2\ j\ m\rangle$ is related to its reduced matrix element relative to the states $|\gamma j_2 m_2\rangle$, where γ includes $j_1\ m_1$:

$$\langle j_1 j_2 j\|\mathbf{V}\|j_1 j_2' j'\rangle = \langle j_2\|\mathbf{V}\|j_2'\rangle(-)^{j+j_1+j_2'+1}(2j+1)^{1/2}$$

$$\times (2j'+1)^{1/2}\begin{Bmatrix} j' & j & 1 \\ j_2 & j_2' & j_1 \end{Bmatrix}$$

where the last factor is a $6j$-symbol. In particular

$$\langle j_1 j_2 j\|\mathbf{J}_2\|j_1 j_2 j'\rangle = [j_2(j_2+1)(2j_2+1)(2j+1)(2j'+1)]^{1/2}$$

$$\times(-)^{j_1+j_2+j+1}\begin{Bmatrix} j' & j & 1 \\ j_2 & j_2 & j_1 \end{Bmatrix} \tag{2.15}$$

This will be used in the next chapter with $j_1 = j_2 = F$. Since $j' = j$ or $j \pm 1$, only two independent $6j$-symbols are required, with explicit forms:

$$\begin{Bmatrix} j & j & 1 \\ F & F & F \end{Bmatrix} = \tfrac{1}{2}(-)^{j+2F+1}\left[\frac{j(j+1)}{(2j+1)F(F+1)(2F+1)}\right]^{1/2} \tag{2.16}$$

$$\begin{Bmatrix} j+1 & j & 1 \\ F & F & F \end{Bmatrix} = \frac{1}{2}(-)^{j+2F+1}\left[\frac{(j+1)(2F+j+2)(2F-j)}{(2j+1)(2j+3)\ F(F+1)(2F+1)}\right]^{1/2} \tag{2.17}$$

It is often convenient to write a vector form of relations between the components of vector operators, especially commutation relations, and between their matrix elements. Thus (2.1) can be written $\mathbf{J} \times \mathbf{J} = i\mathbf{J}$, giving three equations by taking any independent components. A scalar operator S satisfies $[S, \mathbf{J}] = \mathbf{0}$. Many equations in the next chapter use this form and so imply three equations between components. The usual dot and cross products will be used: if $\mathbf{U}$ and $\mathbf{V}$ are vector operators, then $\mathbf{U} \cdot \mathbf{V}$ gives a scalar, and $\mathbf{U} \times \mathbf{V}$ gives a vector, all with respect to the same angular momentum.

Use of the underlying group theory will now be illustrated by considering the effect of changing the signs of m and M in the matrix element (2.12). Changing the sign of the eigenvalues of J_z corresponds to reversing the z-axis, which can be achieved by a rotation through π about the x-axis. If R is the appropriate rotation operator, then $R|jm\rangle$ and $|j\ -m\rangle$ should represent the same physical state. Alternatively, the effect of the rotation can be described by the transformation $A \to RAR^{-1}$ of operators representing observables. Since the rotation also reverses the y-axis, the components of a vector operator should satisfy $RV_x R^{-1} = V_x$, $RV_y R^{-1} = -V_y$, and $RV_z R^{-1} = -V_z$: R commutes with V_x and anticommutes with V_y or V_z.

These results will now be obtained algebraically. $R = \exp(\pi D_{23}) =$

$\exp\left(-i\pi\,L_x\right)$ if $h\mathbf{L} = \mathbf{r} \times \mathbf{p}$; however the algebra follows from (2.1), so $R = \exp\left(-i\pi J_x\right)$ may be used. The work will then apply when j and m are half-integers, in which case the rotation should be interpreted as an SU_2 transformation.

If $A = J_x$ and $B = V_y$ or V_z, the commutators in the expansion (G64)

$$e^{aA}Be^{-aA} = B + a[A, B] + \tfrac{1}{2}a^2[A, [A, B]] + \cdots$$

can written down from (2.10), and $ia = \theta$ gives

$$\exp(-i\theta J_x)\ V_y\ \exp(i\theta J_x) = V_y\cos\theta + V_z\sin\theta$$
$$\exp(-i\theta J_x)\ V_z\ \exp(i\theta J_x) = V_z\cos\theta - V_y\sin\theta$$

These equations give the operator transformations corresponding to a rotation through θ about the x-axis. For $\theta = \pi$ they can be written $RV_{-m} = -V_m R(m = 0, \pm 1)$ using the circular components (2.11).

In particular, taking $\mathbf{V} = \mathbf{J}$, $RJ_\pm = J_\mp R$. Then $J_-(R|j\,m{=}j\rangle) = RJ_+|j\,j\rangle = 0$, and since the unitary operator R does not change the norm of a state, $R|j\,j\rangle \neq 0$. Hence $R|j\,j\rangle = p|j\ -j\rangle$ with $|p| = 1$. Similarly

$$R|j\,j{-}1\rangle = R(2j)^{-1/2}J_-|j\,j\rangle = (2j)^{-1/2}\,J_+R|j\,j\rangle = p|j\ -j{+}1\rangle$$

Continuing thus yields $R|jm\rangle$ for every m ; if the Γ commute with J_x then $R|\gamma jm\rangle = p|\gamma j\ -m\rangle$ where p is a phase factor depending only on j and evidently determined by (2.3).

The phase factor $p = \langle -m|\,R\,|m\rangle$ can be written down from more general results quoted in Chapter 1. The Euler angles of the rotation are $\alpha = \beta = \pi$, $\gamma = 0$, so $p = d^j_{-m\,m}(\pi)\,e^{im\pi}$. From (1.18), $p = (-)^{j-m}e^{im\pi}$ and so $R|jm\rangle = i^{2j}\,|j-m\rangle$.

Then $\langle J\ -M|\,V_{-m}\,|J'\ m-M\rangle = -\langle J\ -M|\,R^\dagger V_m R\,|J'\ m-M\rangle$
$$= i^{2J'-2J+2}\langle JM|\,V_m\,|J'\ M{-}m\rangle$$

Since $J' - J$ is an integer, $i^{2J'-2J+2}$ can be written $(-)^{J'-J+1}$. Substituting the result into (2.12) gives (a special case of) a well-known symmetry property of Clebsch-Gordan coefficients (G64, M58, R57). Cartesian components satisfy $\langle J\ -M|V_\alpha|J'\ -M'\rangle = \pm(-)^{J'-J}\langle JM|V_\alpha|J'M'\rangle$ with the $+$ sign for V_x and the $-$ sign for V_y and V_z.

2.4. INTERPRETATION IN TERMS OF LIE ALGEBRAS

In the language of Section 1.6, equation (2.1) shows that iJ_x, iJ_y, and iJ_z form an o_3 algebra. The representation space consisting of the states of the system gives a skew-Hermitian representation of o_3. In order to allow the J_α

rather than the iJ_α to give the representation, it is convenient to define the Lie product of the J_α as the commutator divided by i. The representation is then Hermitian.

Then Section 2.1 says that the domain is the direct sum of subspaces $\mathscr{S}_{\gamma j}$ giving irreducible representations $[\gamma j]$ labeled by the quantum numbers. The number $(2j + 1)$ is a positive integer and is the dimension of $\mathscr{S}_{\gamma j}$, in which the Casimir operator J^2 has the value $j(j + 1)$ and J_z has the eigenvalues $-j$, $-j + 1, \ldots, j$. The standard basis for $\mathscr{S}_{\gamma j}$, consisting of the $2j + 1$ eigenstates of J_z related by the shift operators according to (2.3), will be called a multiplet. However $J_\pm$ belong not to the o_3 algebra but rather to its complexification.

States with j different are orthogonal since they belong to (the domains of) different irreducible representations of o_3. States with m different are orthogonal because they belong to different irreducible representations of the subalgebra o_2. The function $e^{im\phi}$, with m any integer, gives the domain of an irreducible representation $[m]$ of o_2 or SO_2. Any algebra in which all the Lie products are zero, or correspondingly any Abelian group, has only one-dimensional irreducible representations.

The operators $\exp(-i\alpha J_z)$ and $\exp(-i\beta J_y)$ are unitary when α and β are real. As α and β vary over suitable finite intervals, these operators and their products form a compact group. For an orbital angular momentum a representation of SO_3 is thus obtained while an arbitrary angular momentum leads to a representation of SU_2.

Two commuting angular momenta, with Lie products defined as (commutator)$/i$, give a Hermitian representation of $o_3 \oplus o_3 = o_4 : D_{12} \leftrightarrow J_z = J_{1z} + J_{2z}$, $D_{23} \leftrightarrow J_x$, $D_{31} \leftrightarrow J_y$, $D_{14} \leftrightarrow J_{1x} - J_{2x}$, $D_{24} \leftrightarrow J_{1y} - J_{2y}$, $D_{34} \leftrightarrow J_{1z} - J_{2z}$. The states $|\gamma\, j_1\, m_1\, j_2\, m_2\rangle$ with fixed $\gamma j_1 j_2$ form the domain of an irreducible representation $[\gamma\, j_1\, j_2]$ of o_4, which contains reducible representations of o_3. In particular the $\mathbf{J}_1$ subalgebra gives the direct sum consisting of $[\gamma\, j_1]$ taken $(2j_2 + 1)$ times; the $\mathbf{J}_2$ subalgebra gives the representation $[\gamma\, j_2]$ taken $(2j_1 + 1)$ times; and the subalgebra $\mathbf{J}_1 + \mathbf{J}_2 = \mathbf{J}$ gives the product representation $[j_1] \times [j_2]$. This decomposes into irreducible representations by

$$[\gamma\, |j_1 - j_2|] + [\gamma\, |j_1 - j_2| + 1] + \cdots + [\gamma\, j_1 + j_2],$$

showing the allowed values of the total angular momentum. It is therefore possible to interpret Section 2.2 entirely in terms of the algebra o_3 or the group SU_2, but the application to the Coulomb problem requires the relation to o_4. The Clebsch-Gordan coefficients give the transformation from the standard bases of the representations $[\gamma j_1]$ and $[\gamma j_2]$ to the standard bases of the representations given by the subalgebra $\mathbf{J}$. For irreducible representations of o_2, $[m_1] \times [m_2]$ is $[m_1 + m_2]$ because $\exp(im_1\phi) \exp(im_2\phi) = \exp(im_1\phi + im_2\phi)$. This explains the condition $M = m_1 + m_2$ on the Clebsch-Gordan coefficients.

Scalar and vector operators are representation spaces for o_3 of dimensions

1 and 3 respectively, the representations being the type defined in Section 1.15. A scalar operator S also gives the representation $[0]$ of o_2. Then $S|M'\rangle$ gives the representation $[0] \times [M'] = [M']$ of o_2, the product representation being the kind considered in the third example of Section 1.16. The selection rule $M = M'$ for a nonzero matrix element follows; similarly $J = J'$ is due to $[0] \times [J] = [J]$ for irreducible representations of o_3. Independence of M of the matrix element is also a result of the o_3 symmetry: if R is a representative of the group (SO_3 or SU_2) such that $R|M\rangle = |M'\rangle$, then $\langle M'|S|M'\rangle = \langle M|R^{-1}S\ R|M\rangle$.

A vector operator gives a representation of the subalgebra o_2 ($D_{12} \leftrightarrow J_z$) which is decomposed into irreducible representations $[m]$ by the basis of spherical components V_m. Again the property $[m] \times [M'] = [m + M']$ of o_2 representation gives the selection rule $M = m + M'$ on the matrix elements of the V_m. The $V_m(m=0, \pm 1)$ and the $|\gamma'\ J'\ M-m\rangle$ multiplet ($M=m+J'$, $\dots, m - J'$) give representations of o_3 in both of which the components of $\mathbf{J}$ are the representatives. As in the third example of Section 1.16, their products $V_m|\gamma'\ J'\ M-m\rangle$ give the product representation $[1] \times [J']$, in which the representatives are also the components of $\mathbf{J}$. The decomposition $[1] \times [J'] = [J'-1] + [J'] + [J'+1]$ therefore gives the selection rule on J. Moreover, the basis effecting the decomposition is obtained using the Clebsch-Gordan coefficients, which therefore appear in the matrix elements of the V_m as in the Wigner-Eckart theorem (S68).

The products of the components of two vector operators give the product representation $[1] \times [1]$ in which, as in the second example of Section 1.16, the representatives are still the components of $\mathbf{J}$ acting by taking the commutator. The decomposition $[0] + [1] + [2]$ shows that a scalar and a vector can be formed from the products. These are given by the dot and cross product respectively; using spherical components, the required combinations are given by Clebsch-Gordan coefficients.

2.5. EXTENSION TO AN $o_{2,1}$ ALGEBRA

Suppose that T_x, T_y, and T_z are Hermitian operators with the commutation relations (L66)

$$[T_x, T_y] = -iT_z, \quad [T_y, T_z] = iT_x, \quad [T_z, T_x] = iT_y$$

They form an $o_{2,1}$ algebra: the correspondence $T_z \leftrightarrow D_{12}$, $T_y \leftrightarrow -E_{31}$, $T_x \leftrightarrow E_{23}$ preserves the Lie products in (1.31) if the Lie products of the T_α are defined as (commutator)$/i$. The relations $[T_z, T_x \pm iT_y] = \pm(T_x \pm iT_y)$ show that $T_\pm = T_x \pm iT_y$ are shift operators for eigenstates of T_z which change its eigenvalues by ± 1. The Casimir operator $T^2 = T_x^2 + T_y^2 - T_z^2$ commutes with T_z, so that there are simultaneous eigenstates of T^2 and T_z. Also $[T^2, T_\pm] = 0$, so for a given eigenvalue of T^2, there will be a set of allowed eigenvalues of T_z which differ by integers.

Suppose also that T_z has a minimum eigenvalue k, and denote the corresponding eigenstate by $|k0\rangle$. Then $T_-|k0\rangle = 0$, and as $T^2 = -T_z^2 + T_z + T_+T_-$, $T^2|k0\rangle = k(1 - k)|k0\rangle$. Now using T_+, a sequence $|ka\rangle$ of eigenstates of T_z belonging to the eigenvalues $k + a(a = 0, 1, 2, \ldots)$ can be constructed. It is possible that $T_+|k0\rangle = 0$ also, but if the sequence has more than one term it does not terminate: there is no positive integer n such that $T_+|kn\rangle = 0$. This follows because T_+ and T_- are complex conjugates, and so $T_+|kn\rangle = 0$ implies $0 = \langle kn|T_-T_+|kn\rangle = \langle kn|T^2 + T_z^2 + T_z|kn\rangle$ or $0 = (2k + n)(n + 1)$, $n = -2k$. But $0 \leq \langle kn|T_+T_-|kn\rangle$ giving $0 \leq n(2k + n - 1) = -n$ if $2k = -n$, which is a contradiction. The case $T_\pm|k0\rangle = 0$ occurs when $k = 0$, and otherwise $k > 0$ because $2k + n - 1$ must be positive for $n = 1$.

This argument demonstrates that the only finite-dimensional, irreducible, Hermitian representations of $o_{2,1}$ are one-dimensional.

The analog of (2.3) follows from the equations $\langle ka|T_-T_+|ka\rangle = (2k + a)(a + 1)$ and $\langle ka|T_+T_-|ka\rangle = a(2k + a - 1)$ which imply that if $|ka\rangle$ denotes a normalized state, then

$$T_+|ka\rangle = p(2k + a)^{1/2}(a + 1)^{1/2}|k\ a+1\rangle$$
$$T_-|ka\rangle = \bar{p}\ a^{1/2}(2k + a - 1)^{1/2}|k\ a - 1\rangle \tag{2.18}$$

The phase factor p will depend on a unless a suitable phase is assigned to $|ka\rangle$. In the example of this theory in Section 1.12, if $|ka\rangle$ denotes the normalized function $\left(\dfrac{-b - 1}{a}\right)^{1/2} z^a w^b$, then (1.43) gives (2.18) with $p = -i$. However if $|ka\rangle = \left(\dfrac{-b - 1}{a}\right)^{1/2} i^a z^a w^b$, then (2.18) holds with $p = -1$; while if $|ka\rangle = \left(\dfrac{-b - 1}{a}\right)^{1/2} z^a w^b i^b$, then $p = 1$. Thus the i in (1.47) could be removed by redefining the isometric correspondence.

There are finite-dimensional representations of $o_{2,1}$ for which the Clebsch-Gordan coefficients $\langle j_1\ m_1\ j_2\ m_2|j_1\ j_2\ J\ M\rangle$ give the same results as in Sections 2.2 and 2.3. The decomposition seen in the first example of Section 1.16 is actually $[\frac{1}{2}] \times [1] = [\frac{1}{2}] + [\frac{3}{2}]$. If the representation spaces are states, such applications cannot be physically significant, because the representatives are not Hermitian. Useful results can be obtained when the representation spaces are operators. For example, the nine products in the second example in Section 1.16 include an $o_{2,1}$ scalar and an $o_{2,1}$ vector.

The operators T_x, T_y, T_z defined on the states $|ka\rangle$ give an irreducible representation of $o_{2,1}$ which is isometrically equivalent to that in Section 1.12

through correspondences such as $|ka\rangle \leftrightarrow \left(\dfrac{-b-1}{a}\right)^{1/2} i^a z^a w^b, T_x \leftrightarrow T_1, T_y \leftrightarrow T_2,$
$T_z \leftrightarrow T_3$. Then a result can be written down from (1.46):

$$\langle kA| \exp (i\lambda T_y)|ka\rangle$$

$$= i^{a-A}\left(\frac{a+2k-1}{a}\right)^{1/2}\left(\frac{A+2k-1}{A}\right)^{1/2}\langle BA| \exp (i\lambda T_2)|ba\rangle$$

$$= \frac{(-)^A(a+A+2k-1)!\,(\tanh \tfrac{1}{2}\lambda)^{a+A}\ F}{(\cosh \tfrac{1}{2}\lambda)^{2k}[a!\,A!\,(a+2k-1)!\,(A+2k-1)!\,]^{1/2}} \tag{2.19}$$

where F is the hypergeometric function in (1.46). An alternative form of this result appears by using the transformation formula (M43-54) which expresses a hypergeometric function of argument z in terms of one with argument $1-z$. This gives

$$F = \frac{(a+2k-1)!\,(A+2k-1)!}{(a+A+2k-1)!\,(2k-1)!}\ {}_2F_1(-a,-A;2k;1-\coth^2 \tfrac{1}{2}\lambda) \tag{2.20}$$

If the states $|ka\rangle$ give (2.18) with $p=1$, then the correspondence with Section 1.12 should use $z^a w^b i^b$ instead of $i^a z^a w^b$, and (2.19) will be multiplied by $i^{b-a+A-B} = (-)^{A-a} = (-)^{a\pm A}$.

3. Energy Eigenvalues and Eigenstates

The main physical results in this chapter are an algebraic derivation of the Coulomb energy eigenvalues and their degeneracy and the resulting evaluation of matrix elements of position, momentum, and kinetic energy between states of the same energy. The matrices are obtained relative to two sets of basis states which appear naturally and have group-theoretical significance. Only the angular momentum theory of Chapter 2 is used in these calculations. Another important result is the identification of the invariance algebra and group.

The required algebra is fairly complicated, even though some of the details are hidden by using angular momentum theory and the evaluation of commutators is relegated to an appendix. In case this complication obscures the role played by group theory, I begin with two systems that are simpler from the point of view of the algebraic approach. The first is the two-dimensional oscillator, which not only provides a nice illustration of an invariance group, but is also related to the Coulomb problem in Chapters 5 and 6. The second is the two-dimensional Coulomb problem, which shows both the connection between oscillator and Coulomb problems and the use of eigenvalues of the invariance algebra operators to determine the symmetry group.

However, the treatment of the three-dimensional Coulomb problem in Sections 3.3 through 3.6 does not depend on the treatment of these simpler systems and can be read from Chapter 2 alone.

Section 3.1 on the two-dimensional oscillator assumes the algebraic solution of the one-dimensional oscillator problem. The relation between the two-dimensional oscillator and Coulomb problems is obtained by a nonalgebraic method, namely, comparing their Schrödinger equations.

3.1. THE TWO-DIMENSIONAL HARMONIC OSCILLATOR

If u and v are Cartesian coordinates in a two-dimensional space, a particle of mass μ moving in the harmonic oscillator potential $\frac{1}{2}\mu\omega^2(u^2 + v^2)$ has Hamiltonian $H_u + H_v$, where $2\mu H_u = p_u{}^2 + \mu^2\omega^2 u^2$. Since H_u and H_v commute, the energy eigenstates can be chosen to be eigenstates of H_u and H_v.

These are Hamiltonians of one-dimensional oscillators, with eigenvalues $(s + \frac{1}{2})h\omega$ and $(t + \frac{1}{2})h\omega$, where h is Planck's constant divided by 2π (the constant usually written $\hbar$), and s and t are any nonnegative integers. So the eigenvalues of $H_u + H_v$ are $nh\omega$, where $n = s + t + 1$ is a positive integer. The degeneracy of the eigenvalue $nh\omega$ is n, the number of ways s and t can be chosen so that $s + t = n - 1$. The normalized eigenfunctions ψ_{st} are products of one-dimensional oscillator eigenfunctions: $\psi_{st}(u, v) = \psi_s(u)\psi_t(v)$.

As $A_u = (\mu\omega/2h)^{1/2}u + i(2\mu\omega h)^{-1/2}p_u$ satisfies $[H_u, A_u] = -h\omega A_u$, it is a lowering operator for H_u and H: $A_u\psi_{st}$ is a multiple of ψ_{s-1t}. Similarly A_v is a lowering operator for H_v and H, while A_u^* and A_v^* are raising operators. For the oscillator, * will denote complex conjugate, for reasons given at the end of the next section. The functions ψ_{st} are obviously eigenfunctions of $A_u^* A_u$ and $A_v^* A_v$. In fact s and t are the eigenvalues of $A_u^* A_u$ and $A_v^* A_v$, and the Hamiltonian can be written $(A_u^* A_u + A_v^* A_v + 1)h\omega$. The operator $A_u^* A_v$ increases s and decreases t but leaves $s + t = n$ unaltered, while $A_v^* A_u\psi_{st}$ is a multiple of $\psi_{s-1\ t+1}$. These operators change any energy eigenstate to another of the same energy. In other words, the n-dimensional space of eigenfunctions belonging to the energy eigenvalue $nh\omega$ is invariant under $A_u^* A_v$ and $A_v^* A_u$. These operators must therefore commute with the Hamiltonian, and this can be explicitly verified by using the basic commutators $[A_u, A_u^*] = [A_v, A_v^*] = 1$, $[A_u, A_v] = 0$, etc.

Any Lie algebra containing $A_u^* A_v$ and $A_v^* A_u$ must also include

$$[A_u^* A_v, A_v^* A_u] = A_u^*[A_v, A_v^* A_u] + [A_u^*, A_v^* A_u]A_v$$
$$= A_u^*[A_v, A_v^*]A_u + A_v^*[A_u^*, A_u]A_v = A_u^* A_u - A_v^* A_v$$

The commutators of $A_u^* A_u - A_v^* A_v$ with $A_u^* A_v$ and $A_v^* A_u$ can be calculated in the same way. Alternatively, since ψ_{st} is an eigenfunction of $A_u^* A_u - A_v^* A_v$ belonging to the eigenvalue $s - t$, $A_u^* A_v$ raises this eigenvalue by 2, and $A_v^* A_u$ lowers it by 2. This implies $[A_u^* A_u - A_v^* A_v, A_u^* A_v] = 2A_u^* A_v$ and $[A_u^* A_u - A_v^* A_v, A_v^* A_u] = -2A_v^* A_u$. Thus the three operators are a basis of a Lie algebra.

To identify this algebra with $c*o_3$, let $J_z = \frac{1}{2}(A_u^* A_u - A_v^* A_v)$ so that the eigenvalues of J_z are changed by 1. Then $[J_z, J_\pm] = J_\pm$, $[J_+, J_-] = 2J_z$, with $J_+ = A_u^* A_v$, $J_- = A_v^* A_u$. Angular momentum theory can be applied because J_+ and J_- are complex conjugates. In other words, the Hermitian operators $J_z, \frac{1}{2}(J_+ + J_-)$, and $\frac{1}{2}i(J_- - J_+)$ are a basis of a real o_3 algebra. Since $J_+J_- + J_-J_+ = 2A_u^* A_u A_v^* A_v + A_u^* A_u + A_v^* A_v$, the Casimir operator $J_z^2 + \frac{1}{2}(J_+J_- + J_-J_+)$ is $\frac{1}{4}(A_u^* A_u + A_v^* A_v)^2 + \frac{1}{2}(A_u^* A_u + A_v^* A_v)$.

It is convenient to let J_3 denote $\frac{1}{2}i(J_- - J_+)$, because the physical angular momentum $up_v - vp_u$ is then $2hJ_3$. Then J_z is denoted by J_1, and $\frac{1}{2}(J_+ + J_-)$ by J_2. The solution above of the energy eigenvalue problem assumed the results of the one-dimensional oscillator problem. The existence of an o_3 algebra of operators commuting with the Hamiltonian allows the energy eigenvalue problem to be investigated by angular momentum theory. Given the Hamiltonian, define A_u and A_v as before, and define (J40)

$$J_1 = \tfrac{1}{2}(A_u^* A_u - A_v^* A_v) = (2h\omega)^{-1}(H_u - H_v)$$

$$J_2 = \tfrac{1}{2}(A_u^* A_v + A_v^* A_u) = \frac{\mu\omega}{2h} uv - \frac{h}{2\mu\omega} \frac{\partial^2}{\partial u\,\partial v} \qquad (3.1)$$

$$J_3 = \tfrac{1}{2}i(A_v^* A_u - A_u^* A_v) = \frac{1}{2} i \left(v \frac{\partial}{\partial u} - u \frac{\partial}{\partial v} \right)$$

Then the J_i are the components of an angular momentum, and the possible eigenvalues of $J^2 = J_1^2 + J_2^2 + J_3^2$ are known to be $j(j+1)$ with $j = 0, \tfrac{1}{2}$, $1, \dots$. But the evaluation above of the Casimir operator shows that $J^2 + \tfrac{1}{4} = \tfrac{1}{4}(A_u^* A_u + A_v^* A_v + 1)^2$, and so the possible eigenvalues of the Hamiltonian $(A_u^* A_u + A_v^* A_v + 1)h\omega$ are $\pm(2j+1)h\omega$, or $\pm nh\omega$ where n is a positive integer. The negative values can be rejected because the Hamiltonian is a positive operator, and the existence of all the positive eigenvalues can be shown as in the algebraic treatment of the one-dimensional case by using the shift operators A_u^*, A_v^*, A_u, and A_v.

The states ψ_{st} which appear naturally using Cartesian coordinates are simultaneous eigenstates of J^2 and J_1. The simultaneous eigenstates of J^2 and J_3 are eigenstates of the energy and of the orbital angular momentum $up_v - vp_u$, which is $2hJ_3$ and so has eigenvalues that are *integral* multiples of h. With these states the natural shift operators are defined by (M58) $A_{\pm} = 2^{-1/2}(A_u \mp iA_v)$. Then $A_{\pm}$ lower the energy by $h\omega$; A_- raises the angular momentum by h; A_+ lowers the angular momentum by h; and $A_{\pm}^*$ have the opposite behavior.

The o_3 algebra is an invariance algebra for the two-dimensional harmonic oscillator, because every operator of the algebra commutes with the Hamiltonian. An exponential function on the algebra leads to a group of operators that commute with the Hamiltonian and are called an invariance group, or symmetry group, for the two-dimensional oscillator. From the algebra o_3 either of the groups SO_3 or SU_2 may be obtained. Now the energy eigenvalues correspond to both integral and half-integral values of j, and the invariant spaces of energy eigenstates give irreducible representations of the algebra of every order. The group SO_3 would therefore have double-valued representations, and it is not possible to find coordinates x_1, x_2, x_3 such that the invariance group is just the set of transformations of these coordinates resulting from proper rotations. This suggests that the invariance group is SU_2 rather than SO_3.

An explicit representation of the unitary transformations of the symmetry group is obtained by noticing that $A_u^* A_u + A_v^* A_v$, which apart from irrelevant constants is the Hamiltonian, is a Hermitian form in the variables A_u and A_v. If $[B_u^*\ B_v^*] = [A_u^*\ A_v^*]U$, where U is a 2×2 complex matrix, the Hamiltonian becomes $(B_u^* B_u + B_v^* B_v + 1)\hbar\omega$ provided that the matrix U is unitary. As a function of the shift operators the form of the Hamiltonian is invariant under these transformations, which make up the group U_2. Now, as in Section 1.11, U_2 can be mapped onto an SU_2 subgroup by removing the phase factors $|U|^{1/2}$. The transformation of wave functions associated with an element of U_2 differs from the transformation associated with the corresponding element of SU_2 only by an overall phase factor in the wave functions. Since phase factors have no physical significance it is sufficient to consider the SU_2 group.

Alternatively, the symmetry group can be selected by requiring (D63) that the eigenstates of every energy level induce an irreducible representation of the group (which excludes SO_3) and that every irreducible representation can thus be obtained (which excludes U_2).

The best-known examples of angular momentum operators and eigenstates with half-integral eigenvalues involve spin and hence wave functions with two or more components. It is perhaps worth commenting (J40) that this section provides such examples with a single function ψ_{st} representing each state. Also half-integral eigenvalues only appear when the angular momentum operators are the generators of an SU_2 group rather than a rotation group.

3.2. THE TWO-DIMENSIONAL COULOMB PROBLEM

Cartesian coordinates will now be denoted by x and y; the Coulomb potential for a particle of mass μ can be written $-hP/\mu(x^2 + y^2)^{1/2}$, where P is a constant with the dimensions of momentum. If the potential is caused by attraction between charges $\pm e$, then $P = \mu e^2/h$. To exhibit the invariance algebra it is convenient to use parabolic cylinder coordinates (u, v) defined by $x = \frac{1}{2}(u^2 - v^2)$, $y = uv$. Then $(x^2 + y^2)^{1/2} = \frac{1}{2}(u^2 + v^2)$, and

$$\frac{\partial^2}{\partial x^2} + \frac{\partial^2}{\partial y^2} = (u^2 + v^2)^{-1}\left(\frac{\partial^2}{\partial u^2} + \frac{\partial^2}{\partial v^2}\right)$$

Multiplying the Schrödinger equation through by $(u^2 + v^2)$ therefore gives

$$-\frac{h^2}{2\mu}\left(\frac{\partial^2\psi}{\partial u^2} + \frac{\partial^2\psi}{\partial v^2}\right) - \frac{2hP}{\mu}\,\psi = E(u^2 + v^2)\psi$$

where $\psi(u, v)$ is an energy eigenfunction belonging to the eigenvalue E. As $(-u, -v)$ and (u, v) represent the same point of the plane, wave functions must satisfy (C69) $\psi(u, v) \equiv \psi(-u, -v)$. Comparison with the Schrödinger equation of the two-dimensional oscillator of the preceding section

$$-\frac{h^2}{2\mu}\left(\frac{\partial^2\psi}{\partial u^2} + \frac{\partial^2\psi}{\partial v^2}\right) + \frac{1}{2}\mu\omega^2(u^2 + v^2)\psi = E'\psi$$

shows that using parabolic cylinder coordinates has effectively converted the Coulomb problem to the oscillator problem. The oscillator eigenfunction $\psi_{st}(u, v)$, belonging to the oscillator eigenvalue $E' = nh\omega = (s + t + 1)h\omega$, is an eigenfunction of the Coulomb problem provided that $E' = 2hP/\mu$ and $\psi_{st}(u, v) = \psi_{st}(-u, -v)$. Now

$$\psi_{st}(-u, -v) = \psi_s(-u)\psi_t(-v) = (-)^{s+t}\psi_s(u)\psi_t(v),$$

since the one-dimensional oscillator wave functions are odd or even according as s (or t) is odd or even. Thus $s + t$ must be even, and n must be odd. For any odd integer n, an oscillator can be chosen with its constant ω satisfying $\mu n\omega = 2P$, and then the n independent eigenfunctions $\psi_{st}(s + t = n - 1)$ are eigenfunctions of the Coulomb problem belonging to the Coulomb eigenvalue $E = -\frac{1}{2}\mu\omega^2 = -2P^2/\mu n^2$. Putting $n = 2j + 1$, the eigenvalues are $-2P^2/\mu(2j + 1)^2$ with $j = 0, 1, 2, \ldots$, and the degeneracy is $(2j + 1)$. This agrees with results obtained (Z67) by solving the Schrödinger equation in plane polar coordinates.

Using properties of the oscillator to deduce other facts about the Coulomb system is complicated by ω depending on n, hence on the Coulomb eigenvalue. The comparison above of the oscillator in Cartesian coordinates and the Coulomb system in parabolic coordinates gives a mapping of Coulomb eigenfunctions to oscillator eigenfunctions, but the states of different Coulomb energy levels correspond to states of oscillators with different constants. However, the degenerate states belonging to one Coulomb level correspond to the degenerate states of one level of one oscillator, and so the invariance algebra of the oscillator is an invariance algebra for this energy level of the Coulomb problem.

A basis for this o_3 invariance algebra is given by (3.1), with the domain of the operators the n-dimensional space of energy eigenfunctions belonging to an eigenvalue E, and $\omega = (-2E/\mu)^{1/2}$. Every function of the domain is an eigenfunction of H belonging to the eigenvalue E, so the constant $\mu\omega^2$ can be replaced by $-2H$ where convenient, provided that the operator $-2H$ is placed to the right of any other operators. When the operators (3.1) are expressed in the Cartesian coordinates (x, y), the results are simplified if this device is used to write $\mu\omega^2(u^2 - v^2)$ in J_1 as $-2(u^2 - v^2)H$ and to write $\mu\omega uv$ in J_2 as $-2\omega^{-1}uvH$. Then

$$J_1 = \frac{1}{2\mu\omega}\left[-Lp_y - p_yL + \frac{2xP}{(x^2 + y^2)^{1/2}}\right]$$

$$J_2 = \frac{1}{2\mu\omega}\left[Lp_x + p_xL + \frac{2yP}{(x^2 + y^2)^{1/2}}\right] \qquad (3.2)$$

$$J_3 = L = \frac{1}{h}(xp_y - yp_x)$$

with $\omega = (-2E/\mu)^{1/2}$. Also $H_u + H_v = E' = 2hP/\mu$, so that $A_u^* A_u + A_v^* A_v + 1 = 2P/\mu\omega$, and from the preceding section $J^2 + \tfrac{1}{4} = P^2/\mu^2\omega^2$, which can also be got directly from (3.2) if $2H$ is replaced by $-\mu\omega^2$ on reaching $J_1{}^2 + J_2{}^2 = (\mu\omega)^{-2}[P^2 + 2\mu(L^2 + \tfrac{1}{4})H]$.

The condition $\psi(u, v) \equiv \psi(-u, -v)$ restricts the quantum number $j = \tfrac{1}{2}n - \tfrac{1}{2}$ to integral values. As $j = \tfrac{1}{2}$ is excluded, the 2×2 matrix representation of the algebra cannot be obtained, and the corresponding group does not have a representation by 2×2 unitary matrices. The symmetry group (C69) is SO_3, and coordinates x_1, x_2, x_3 should exist in which (3.2) become $iD_{\alpha\beta}$, defined in (1.7), and in which the group is the set of transformations obtained from proper rotations.

The operators (3.2) are not identical to (3.1), because the domains are different. Also in the oscillator problem, $(\partial/\partial u)^* = -(\partial/\partial u)$, because the area element is $du\, dv$. In the Coulomb problem the area element is $(u^2 + v^2)\, du\, dv$, so that $-(\partial/\partial u)$ is not the complex conjugate of $(\partial/\partial u)$. So complex conjugates are denoted with $*$ for the oscillator, but with $\dagger$ in the Coulomb problem.

Having determined operators that are a standard basis of an o_3 invariance algebra, an algebraic treatment of the energy eigenvalue problem can be formulated. Only angular momentum theory is used, and the relation to the two-dimensional oscillator is not required. Given H, suppose that E is an eigenvalue and define the J_i by (3.2) (with $\omega^2 = -2E/\mu$) with domain the eigenfunctions belonging to E. The commutators of an angular momentum can be verified, and $J_i^\dagger = J_i$ if $E < 0$. The known eigenvalues of J^2 give $P^2/\mu^2\omega^2 = (j + \tfrac{1}{2})^2$ and hence $-2P^2/\mu(2j + 1)^2$ as the possible negative eigenvalues ($j = 0, \tfrac{1}{2}, 1, \ldots$). However, a nonalgebraic argument is required to exclude the half-integral values of j; for example, J_3 must have integral eigenvalues, since half-integral eigenvalues for the orbital angular momentum imply double-valued wave functions.

3.3. THE THREE-DIMENSIONAL COULOMB PROBLEM

The procedure described in the preceding paragraph will now be applied to the three-dimensional problem to give an algebraic determination of the eigenvalues. Six constants of the motion are known, and their commutators show that an o_4 invariance algebra can be defined, provided that the domain is restricted to the states belonging to one energy level. The argument can be given in terms of angular momentum theory because $o_4 = o_3 \dotplus o_3$. The required commutators are evaluated in Appendix A.

The Hamiltonian H for the relative motion in the hydrogenlike atom is given by $\mu H = \tfrac{1}{2}p^2 - hPr^{-1}$, where μ is the reduced mass, P is a constant with the dimensions of momentum, and h is Planck's constant divided by 2π. The Bohr radius is then h/P.

The Hamiltonian is spherically symmetric and so commutes with the angular momentum operators, the components of $h\mathbf{L} = \mathbf{r} \times \mathbf{p}$. The components of the vector (P26-67)

$$\mathbf{A} = \tfrac{1}{2}(\mathbf{p} \times \mathbf{L}) - \tfrac{1}{2}(\mathbf{L} \times \mathbf{p}) - Pr^{-1}\mathbf{r}$$

or   (3.3)

$$h\mathbf{A} = \tfrac{1}{4}\mathbf{r}p^2 + \tfrac{1}{4}p^2\mathbf{r} - \tfrac{1}{2}(\mathbf{r} \cdot \mathbf{p})\mathbf{p} - \tfrac{1}{2}(\mathbf{p} \cdot \mathbf{r})\mathbf{p} + \mu\mathbf{r}H$$

also commute with H. These constants of the motion are less familiar than the components of $\mathbf{L}$ and so are sometimes called hidden. The associated symmetry, which will be elucidated in this chapter and the next, is certainly not obvious. $\mathbf{A}$ was discovered in the classical mechanics problem. However, it can be made to appear by solving the Schrödinger equation in a suitable way, for example, in parabolic coordinates, by factorization, or in momentum space. If $p_z = 0$ and $L_z = L$, the components A_x and A_y are the terms in square brackets in (3.2), apart from a sign change.

Since $\mathbf{A}$ is a vector, the components of $\mathbf{L}$ and $\mathbf{A}$ have the commutators (2.10), which imply

$$\mathbf{A} \cdot \mathbf{L} = \mathbf{L} \cdot \mathbf{A}, \; \mathbf{L} \times \mathbf{A} + \mathbf{A} \times \mathbf{L} = 2i\mathbf{A}$$

From (3.3) one can also show (P26-67)

$$\mathbf{A} \cdot \mathbf{L} = 0, \; \mathbf{A}^2 = P^2 + 2\mu H(\mathbf{L}^2 + 1), \; \mathbf{A} \times \mathbf{A} = -2\mu i H \mathbf{L} \qquad (3.4)$$

These results are derived in Appendix A. Components of the last equation are $[A_x, A_y] = i(-2\mu H)L_z$, etc. Comparison with (1.10) shows that if $-2\mu H$ is replaced by a constant, say $p_0{}^2$, and $\mathbf{M} = \mathbf{A}/p_0$, then the components of $\mathbf{L}$ and $\mathbf{M}$ give an o_4 algebra (K33). The alternative basis (1.11) enables the treatment below to be given in terms of angular momentum theory.

For any number q, the linear combinations $\mathbf{J}_1 = \tfrac{1}{2}\mathbf{L} + q\mathbf{A}, \mathbf{J}_2 = \tfrac{1}{2}\mathbf{L} - q\mathbf{A}$ are alternative constants of the motion satisfying ($k = 1, 2$):

$$\mathbf{J}_k{}^2 = \tfrac{1}{4}\mathbf{L}^2 + q^2\mathbf{A}^2 = q^2(2\mu H + P^2) + \tfrac{1}{4}\mathbf{L}^2(1 + 8\mu q^2 H) \qquad (3.5)$$

$$[J_{1x}, J_{2x}] = 0, \; [J_{1x}, J_{2y}] = \tfrac{1}{4}i L_z(1 + 8\mu q^2 H), \ldots \qquad (3.6)$$

$$\mathbf{J}_k \times \mathbf{J}_k = i\mathbf{J}_k - \tfrac{1}{4}i\mathbf{L}(1 + 8\mu q^2 H) \qquad (3.7)$$

If the domain of these operators can be chosen so that $1 + 8\mu q^2 H = 0$, and if q is real, equations (3.6) and (3.7) say that the $\mathbf{J}_i$ are commuting angular momenta. This leads to a simple algebraic derivation of the energy eigenvalues (H33).

Suppose that $E < 0$ is an eigenvalue of H and let $\mathscr{S}$ be the linear space spanned by the eigenstates belonging to E. Equations (3.5) through (3.7) suggest that q should be chosen so that $8\mu q^2 E = -1$. So in $\mathscr{S}$ define $A_n = (-8\mu E)^{-1/2}A$, $F = \frac{1}{2}L - A_n$, and $G = \frac{1}{2}L + A_n$. Then each component of F commutes with any component of G and

$$F^2 = G^2 = -\frac{1}{4} - \frac{P^2}{8\mu E}, \quad F \times F = iF, \quad G \times G = iG. \tag{3.8}$$

Since F is an angular momentum, the possible eigenvalues of F^2 are $F(F+1)$, with $F = 0, \frac{1}{2}, 1, \dots$. But F^2 is constant, so just one of these values of F occurs, and

$$-\frac{1}{4} - \frac{P^2}{8\mu E} = F(F+1)$$

giving

$$E = -\frac{P^2}{8\mu(F + \frac{1}{2})^2} = -\frac{P^2}{2\mu(2F + 1)^2} = -\frac{P^2}{2\mu n^2}$$

where n is the positive integer called the principal quantum number. Where convenient, E, $\mathscr{S}$, F, G, and L may be labeled with a subscript n; on an operator, the subscript implies the domain. Strictly speaking, L_n should be used instead of L, which is an extension of each L_n. Readers who relish such points may wish to add the subscripts, although they have probably already abandoned this book in disgust!

Every state of $\mathscr{S}$ is an eigenstate of F^2 and G^2 belonging to the eigenvalue $\frac{1}{4}(n^2 - 1)$, and a basis for $\mathscr{S}$ can be chosen from simultaneous eigenstates of F_z and G_z, which commute and are constants of the motion. Since the raising and lowering operators $F_\pm$, $G_\pm$ also commute with H, using them cannot give a state not in $\mathscr{S}$. Hence from a given state $|nfg\rangle$ belonging to the eigenvalues f and g of F_z and G_z, the usual $2F + 1 = n$ eigenstates of F_z can be obtained, and the usual $2F + 1$ eigenstates of G_z. It follows that the degeneracy is a multiple of n^2.

This argument only reveals the possible eigenvalues and does not show that eigenstates exist for all n (or even for any n), because $\mathscr{S}_n$ could consist of the zero element only. The wave functions found in the next two chapters show that all values of n occur.

Since $A_n \times A_n = \frac{1}{4}iL$, the commutators of the components of L and $2A_n$ are just equations (1.10), and so (K33) L_x, A_{nx}, etc., are a basis for an o_4 algebra. F and G correspond to the basis (1.11) and are generators of commuting SU_2 subgroups. However, the invariance group is not $SU_2 \times SU_2$, because the Casimir operators F^2 and G^2 of these subgroups have the same value, which is not true for all representations of $SU_2 \times SU_2$. Comparison with the result $K^2 = N^2$ in Section 1.2 shows (B67) that the invariance group is SO_4. A suitable choice of coordinates will give the components of **L** and **A** the form (1.7). Four coordinates are required, and in these coordinates H will have four-dimensional spherical symmetry. The details appear in the next chapter.

The components of any two commuting angular momenta $\mathbf{J}_1$ and $\mathbf{J}_2$ generate an SO_4 group if $\mathbf{J}_1{}^2 = \mathbf{J}_2{}^2$; for example, the spin components of two particles each with spin s. Conversely, a four-dimensional rotational symmetry may be investigated using angular momentum theory. The methods of the next three sections only depend on results from Chapter 2.

3.4. BASES FOR ENERGY EIGENSTATES

If spin is not considered, the states $|nfg\rangle$ are not degenerate, and the n^2 states obtained by taking all values of f and g ($f, g = \frac{1}{2} - \frac{1}{2}n, \frac{3}{2} - \frac{1}{2}n, \ldots, \frac{1}{2}n - \frac{1}{2}$) are a basis for $\mathscr{S}$. If the states $|nfg\rangle$ are degenerate, and the additional quantum numbers γ remove the degeneracy, then each value of γ defines a subspace of $\mathscr{S}$, and the n^2 states $|\gamma nfg\rangle$ are a basis for this subspace.

From the theory of addition of commuting angular momenta described in Section 2.2, an alternative basis consists of simultaneous eigenstates of $(\mathbf{F} + \mathbf{G})^2$ and $F_z + G_z$. Since $\mathbf{F} + \mathbf{G} = \mathbf{L}$, this basis is just the orbital angular momentum eigenstates which are usually given in elementary treatments of the atom. Denoting them by $|nlm\rangle$, they are related to the $|nfg\rangle$ through the Clebsch-Gordan coefficients:

$$|nlm\rangle = \sum_{fg} |nfg\rangle\langle F f F g | F F l m = f+g\rangle \qquad (3.9)$$

For given n, the allowed values of l are $0, 1, \ldots, 2F = n - 1$.

In Chapter 5 the wave functions corresponding to these states will be given. For this the following remark is very important: the states $|nfg\rangle$ and $|nlm\rangle$ are normalized and their relative phase is completely specified by the conventions of Chapter 2. The $|nfg\rangle$ satisfy (2.3), and the $|nlm\rangle$ are then given by (3.9), in which the Clebsch-Gordan coefficients have phases specified in (2.6). Thus once one state $|nfg\rangle$ of $\mathscr{S}$ is identified with a particular wave function, there is no further arbitrariness in the wave functions comprising the basis, and care must be taken to assign the wave functions the appropriate phase factor determined by (2.3) and (2.6).

3.5. PARITY

The object of this section is to treat parity algebraically as far as possible. Suppose that Π is a Hermitian operator satisfying the six anticommutation relations $\Pi r = -r\Pi$, $\Pi p = -p\Pi$. Then Π^2 commutes with any component of r or p, and is therefore a number c, which is positive since $\Pi^2 = \Pi^\dagger \Pi$. Redefining Π by multiplying by $c^{-1/2}$ makes Π unitary, and its possible eigenvalues are ± 1. Eigenstates of the parity operator Π belonging to the eigenvalue $+1(-1)$ are called even (odd) parity states.

As Π commutes with H, L^2, and L_z, the states $|nlm\rangle$ are eigenstates of Π if they are nondegenerate, and can be chosen so if they are degenerate. Also Π commutes with $L_\pm$, so the $2l + 1$ states of fixed nl have the same parity.

Next consider the $|nfg\rangle$. The state $|n\ f=g=F\rangle$ is $|n\ l=m=n-1\rangle$ and therefore has definite parity: $\Pi|nFF\rangle = \pm|nFF\rangle$. Since Π commutes with L and anticommutes with A_n, $\Pi F = G\Pi$ and $\Pi G = F\Pi$. Hence, with k a number determined by (2.3),

$$\Pi|nfg\rangle = \Pi k(F_-)^{F-f}(G_-)^{F-g}|nFF\rangle$$
$$= k(G_-)^{F-f}(F_-)^{F-g}\Pi|nFF\rangle = \pm|ngf\rangle$$

Putting this result into (3.9), and using (2.8),

$$\Pi|nlm\rangle = \sum_{fg} \pm|ngf\rangle\langle FgFf\ |FFlm\rangle(-)^{2F-l}$$
$$= \pm(-)^{n-1-l}\ |nlm\rangle$$

Since $\pm$ depends only on n, this shows that parity of $|nlm\rangle$ depends on l through the factor $(-)^l$.

The further assumption that all $l = 0$ states have even parity implies that $\pm = (-)^{n-1}$, so that

$$\Pi|nfg\rangle = (-)^{n-1}|ngf\rangle \quad \text{and} \quad \Pi|nlm\rangle = (-)^l|nlm\rangle \quad (3.10)$$

3.6. MATRIX ELEMENTS

The matrix elements of r are physically important in the calculation of transition probabilities and in the perturbation treatment of the Stark effect. Pauli (P26-67) gave an algebraic derivation of the matrix elements between states of $\mathscr{S}$, that is, between degenerate states belonging to the same energy. This method may be presented (F66) as an application of angular momentum theory.

Matrix elements of r between states of the same parity are zero because r anticommutes with the parity operator Π:

$$\langle 1+|r|2+\rangle = \langle 1+|\Pi r|2+\rangle, \quad \text{since} \quad \Pi = \Pi^\dagger$$
$$= -\langle 1+|r\Pi|2+\rangle = -\langle 1+|r|2+\rangle$$

where + indicates an even parity state and the use of **r** signifies the equation true for any component. Evidently **r** can be replaced by **p** or **A**. Similarly, using (3.10) gives (for **r**, **p**, or **A**)

$$\langle nlm|\mathbf{r}|n'l'm'\rangle = (-)^{l-l'+1}\langle nlm|\mathbf{r}|n'l'm'\rangle$$

$$\langle nfg|\mathbf{r}|n'f'g'\rangle = (-)^{n-n'+1}\langle ngf|\mathbf{r}|n'g'f'\rangle$$

Between states of the same negative energy, the matrix element of any commutator $[B, H]$ is zero:

$$\langle n|BH - HB|n\rangle = \langle n|BH|n\rangle - \langle n|HB|n\rangle = E\langle n|B|n\rangle - E\langle n|B|n\rangle$$

since $H^\dagger = H$. For example, the results of Appendix A give

$$(\mu\mathbf{r}\cdot\mathbf{p})H - H(\mu\mathbf{r}\cdot\mathbf{p}) = ih(\mu H + \tfrac{1}{2}p^2) \tag{3.11}$$

hence $0 = \langle nlm|ih(\mu H + \tfrac{1}{2}p^2)|nl'm'\rangle$, and so

$$\langle nlm|p^2|nl'm'\rangle = -2\mu E_n\delta_{l,l'}\,\delta_{m,m'},$$

which is the virial theorem. Similarly

$$[\mu\mathbf{r}, H] = ih\mathbf{p}, [\mu r^2, H] = ih(\mathbf{r}\cdot\mathbf{p} + \mathbf{p}\cdot\mathbf{r}), [\mu\mathbf{p}, H] = -ih^2 P\mathbf{r}r^{-3} \tag{3.12}$$

lead respectively to $\langle nlm|\mathbf{p}|nl'm'\rangle = 0$, $\langle nlm|\mathbf{r}\cdot\mathbf{p}|nl'm'\rangle = \tfrac{3}{2}ih\delta_{l,\,l'}\,\delta_{m,\,m'}$, and $\langle nlm|\mathbf{r}r^{-3}|nl'm'\rangle = \mathbf{0}$. Taking matrix elements of these commutators between states of different energy leads to results such as $\langle nlm|ih\mathbf{p}|n'l'm'\rangle = \mu(E_{n'} - E_n)\langle nlm|\mathbf{r}|n'l'm'\rangle$, which are known as hypervirial theorems. Other results are obtained by considering double commutators (BB65, S70).

Angular momentum theory will now be used to find the matrix elements of **A**. Since

$$A_z = (-8\mu E)^{1/2}A_{nz} = (-2\mu E)^{1/2}(G_z - F_z), A_z|nfg\rangle = n^{-1}P(g-f)|nfg\rangle.$$

Thus A_z is diagonal relative to the basis $|nfg\rangle$, and $\langle nfg|A_z|n'f'g'\rangle = n^{-1}P(g-f)\delta_{n,\,n'}\,\delta_{f,\,f'}\,\delta_{g,\,g'}$. To obtain the matrix elements relative to the $|nlm\rangle$, it is sufficient to get the reduced matrix element $\langle nl\|\mathbf{A}\|n'l'\rangle$ of (2.12), which contains $\delta_{n,\,n'}$ since **A** is a constant of the motion. Then $\langle nl\|\mathbf{A}\|nl'\rangle = n^{-1}P(\langle nl\|2\mathbf{G}\|nl'\rangle - \langle nl\|\mathbf{L}\|nl'\rangle)$, and the reduced matrix elements of **G** and **L** are given by (2.15) and (2.13). Hence

$$\langle nl\|\mathbf{A}\|nl'\rangle = (-)^{n+l}P\begin{Bmatrix} l' & l & 1 \\ F & F & F \end{Bmatrix}\left[\frac{n^2 - 1)(2l'+ 1)(2l + 1)}{n}\right]^{1/2}$$

$$- \frac{P}{n}\delta_{l,\,l'}[l(l + 1)(2l + 1)]^{1/2} \tag{3.13}$$

This is equivalent to a special case of a result given by Biedenharn (B61) in a purely group-theoretic context. Putting $l = l'$, and substituting from (2.16) the explicit form of the $6j$-symbol, gives $\langle nl\|A\|nl\rangle = 0$, as required by parity considerations.

The matrix elements of $\mathbf{r}$ between states of the same energy are related to those of A by a hypervirial theorem. For if $B = -2\mathbf{r} + (ih)^{-1}\{\mathbf{r}(\mathbf{r} \cdot \mathbf{p}) - 2r^2\mathbf{p}\}$ then (3.11) and (3.12) give $[\mu B, H] = 3h A - 4\mu \mathbf{r}H$. Taking matrix elements between states of the same energy yields

$$4\mu E_n\langle nl\|\mathbf{r}\|nl'\rangle = 3h\langle nl\|A\|nl'\rangle.$$

Therefore

$$\langle nl\|\mathbf{r}\|n\,l + 1\rangle = \tfrac{3}{2}hP^{-1}n(-)^{1+n+l}[n(n^2 - 1)(2l + 1)(2l + 3)]^{1/2}$$

$$\times \begin{Bmatrix} l + 1 & l & 1 \\ F & F & F \end{Bmatrix} \tag{3.14}$$

$$= -\tfrac{3}{2}hP^{-1}n[(l + 1)(n + l + 1)(n - l - 1)]^{1/2}$$

on substituting from (2.17). Now the matrix elements of z and $\mp 2^{-1/2}(x \pm iy)$, the circular components of $\mathbf{r}$, can be written down from the Wigner-Eckart theorem (2.12).

With respect to the states $|nfg\rangle$, the matrix elements of components of F and G are given by (2.3), and those of $A = n^{-1}P(G - F)$ can be written down. The same hypervirial theorem then gives the $\langle nfg|\mathbf{r}|nf'g'\rangle$. The nonzero cases are

$$\langle nfg|z|nfg\rangle = -\frac{3nh}{2P}(g - f)$$

$$\langle nfg|x \pm iy|n\,f \mp 1\,g\rangle = \frac{3nh}{4P}[n^2 - 1 - 4f^2 \pm 4f]^{1/2} \tag{3.15}$$

$$\langle nfg|x \pm iy|n\,f\,g \mp 1\rangle = -\frac{3nh}{4P}[n^2 - 1 - 4g^2 \pm 4g]^{1/2}$$

The obvious relation between the last two expressions has already been derived at the beginning of the section by considering parity.

3.7. GROUP-THEORETICAL COMMENTS

A representation of o_4 may be specified (B61) by the notation $[p, q]$ where $p^2 + q^2 + 2p$ and $pq + q$ are the eigenvalues of the Casimir operators $L^2 + M^2$ and $L \cdot M$. The n^2-dimensional space of degenerate energy eigenstates belonging to the eigenvalue $-P^2/2\mu n^2$ is the domain of a Hermitian $[n - 1, 0]$ representation, the representatives of the standard basis of the algebra being the components of $\mathbf{L}$ and $2\mathbf{A}_n = \mathbf{A}(n/P)$. In this representation the Casimir operator $\mathbf{L} \cdot \mathbf{A}$ is zero, and so the group generated is SO_4. The representative of any element or Casimir operator of the algebra commutes with the Hamiltonian and is thus a constant of the motion. However, there are two standard choices for complete sets of commuting observables. Taking L^2 and L_z, which are Casimir operators of o_3 and o_2 subalgebras generating SO_3 and SO_2 subgroups, gives the basis $|nlm\rangle$. The representation of the SO_3 subgroup generated by $\mathbf{L}$ is decomposed into irreducible representations $[0], [1], \ldots, [n-1]$ by using this basis. On the other hand, taking F_z and G_z gives the basis $|nfg\rangle$ and irreducible representations $[F]$ of the commuting SU_2 subgroups generated by $\mathbf{F}$ and $\mathbf{G}$.

The matrix elements of $\mathbf{A}$ can be written down as in (3.13) because the components of $\mathbf{A}$ are representatives of elements of the o_4 algebra.

The components of $\mathbf{F}$ (or $\mathbf{G}$) also span a representation space of the algebra o_4, the representatives still being the components of $\mathbf{L}$ and $2\mathbf{A}_n$, but now acting by taking the commutator as in Section 1.15. In this representation $\mathbf{F}^2 \neq \mathbf{G}^2$. Since $\mathbf{G}$ and $\mathbf{F}$ commute, $\mathbf{G}^2$ maps any component of $\mathbf{F}$ into zero. To find the action of $\mathbf{F}^2$, consider F_x^2. The action of F_x is $F_x \to 0$, $F_y \to iF_z$, $F_z \to -iF_y$, hence the action of F_x^2 is $F_x \to 0 \to 0$, $F_y \to iF_z \to F_y$, and $F_z \to -iF_y \to F_z$. Generally F_α^2 gives $F_\alpha \to 0$ and $F_\beta \to F_\beta$ ($\beta \neq \alpha$; α, $\beta = x$, y, or z), and so $\mathbf{F}^2$ maps F_β into $2F_\beta$. Thus $\mathbf{F}^2 = 2$ and $\mathbf{G}^2 = 0$; $\mathbf{F}$ is the domain of a $[1, 1]$ representation. (Section 4.3 shows that $\mathbf{M} \leftrightarrow -2\mathbf{A}_n = \mathbf{F} - \mathbf{G}$).

4. Momentum Space

In Chapter 3 the algebraic treatment of energy eigenstates used operators that were a basis for an o_4 algebra and the generators of an SO_4 group; hence there should be some coordinate system in which the problem displays four-dimensional rotational invariance. These coordinates were discovered by Fock (F35) without reference to the algebraic method; the connection was pointed out by Bargmann (B36). Since the generators $\mathbf{A}_n$ of the group are defined differently in each subspace $\mathscr{S}_n$, the required coordinates will also depend on n. Fock's coordinates are momenta obtained by projecting the three-dimensional momentum space onto a four-dimensional sphere with radius depending on n.

This method works in any number of dimensions (A57)—the N-dimensional problem can be solved in momentum space by projecting onto an $(N + 1)$-dimensional sphere. The easiest way to visualize the procedure is to first consider the two-dimensional problem (S65). Section 3.2 showed that its invariance group is the three-dimensional rotation group. In Section 4.2 the momentum space Schrödinger equation is expressed in a form invariant under rotations of a sphere obtained from the physical momentum space by projection. The solutions are shown to be spherical harmonics by using the addition theorem of Section 1.8. In the three-dimensional coordinate system on the sphere the generators are the orbital angular momentum components appearing in (1.9) or (1.35); transforming back to the physical coordinates gives the operators (3.2) used in Section 3.2.

Section 4.3 gives the analogous work for the three-dimensional problem. The solutions of the equation obtained in four-dimensional hyperspace are the hyperspherical harmonics of Section 1.3, with the Y_{nlm} corresponding to the $|nlm\rangle$ of Chapter 3, and the $Z_{n_f g}$ corresponding to the $|nfg\rangle$.

An inexplicable degeneracy of the energy eigenvalues points to the existence of a hidden symmetry. Another indicator is the separability of the Schrödinger equation in more than one coordinate system. For example, the rotational symmetry of two-dimensional central potentials implies that the Schrödinger equation separates in polar coordinates; a hidden symmetry in the oscillator and Coulomb problems is suggested by the fact that separation is also possible in Cartesian and parabolic cylinder coordinates respectively. Conversely, the discovery of a symmetry may determine new coordinate systems in which the

59

Schrödinger equation is separable. This will be illustrated in Section 4.4. The separation process is more familiar for a differential equation than for the usual integral equation used in momentum space. So, using a method due to Hylleraas (H32), the momentum space Schrödinger equation will be written as a differential equation.

4.1. TRANSFORMATION TO MOMENTUM SPACE

If a state is represented by the wave function $\psi(\mathbf{x})$, where $\mathbf{x}$ stands for the Cartesian coordinates of position $x_1, x_2, \ldots, x_N$, then the momentum space wave function for the state is (again writing h instead of $\hbar$)

$$\Psi(\mathbf{p}) = (2\pi h)^{-N/2} \int_{-\infty}^{\infty} \int_{-\infty}^{\infty} \cdots \int_{-\infty}^{\infty} dx_1\, dx_2 \cdots dx_N\, \psi(\mathbf{x}) \exp\left(-ih^{-1}\mathbf{x} \cdot \mathbf{p}\right)$$

(4.1)

where $\mathbf{p}$ stands for the Cartesian components of momentum $p_1, \ldots, p_N$ and $\mathbf{x} \cdot \mathbf{p} = \sum_i x_i p_i$. Then momentum components are represented by multiplicative operators, and the ith position coordinate is represented by $ih(\partial/\partial p_i)$. Substituting these operators into the Hamiltonian will lead to an equation for energy eigenfunctions, but it is more usual to represent the potential energy by an integral operator. Using the inverse of (4.1), the momentum space wave function corresponding to $V(\mathbf{x})\Psi(\mathbf{x})$ may be written

$$\int_{-\infty}^{\infty} \cdots \int_{-\infty}^{\infty} dq_1 \cdots dq_N\, K(\mathbf{p}, \mathbf{q})\Psi(\mathbf{q})$$

with

$$K(\mathbf{p}, \mathbf{q}) = (2\pi h)^{-N} \int_{-\infty}^{\infty} \cdots \int_{-\infty}^{\infty} dx_1 \cdots dx_N\, V(\mathbf{x}) \exp\left[ih^{-1}\mathbf{x} \cdot (\mathbf{q} - \mathbf{p})\right]$$

(4.2)

The Coulomb potential shape $V(\mathbf{x}) = |\mathbf{x}|^{-1} = (x_1^2 + \cdots + x_N^2)^{-1/2}$ gives (S65, K62)

$$K(\mathbf{p}, \mathbf{q}) = \begin{cases} (2\pi h\, |\mathbf{p} - \mathbf{q}|)^{-1} & (N = 2) \\ (2\pi^2 h\, |\mathbf{p} - \mathbf{q}|^2)^{-1} & (N = 3) \end{cases}$$

(4.3)

The integral equation for the energy eigenfunctions of the three-dimensional problem is therefore

$$p^2\Psi(\mathbf{p}) - \frac{P}{\pi^2} \int_{-\infty}^{\infty} \int_{-\infty}^{\infty} \int_{-\infty}^{\infty} dq_x\, dq_y\, dq_z |\mathbf{p} - \mathbf{q}|^{-2}\Psi(\mathbf{q}) = 2\mu E\Psi(\mathbf{p})$$

(4.4)

using q_x, q_y, q_z for q_1, q_2, q_3. A direct solution by separating in spherical polar coordinates was obtained by Eriksen (E62).

4.2.   THE TWO-DIMENSIONAL CASE

For $N = 2$, (4.3) gives the integral equation

$$(p_x^2 + p_y^2 - 2\mu E)\Psi(p_x, p_y) = \frac{P}{\pi} \int_{-\infty}^{\infty} \int_{-\infty}^{\infty} dq_x \, dq_y \, \frac{\Psi(q_x, q_y)}{|\mathbf{p} - \mathbf{q}|} \qquad (4.5)$$

which becomes

$$(p^2 - 2\mu E)\Psi(p, \phi) = \frac{P}{\pi} \int_0^{\infty} q \, dq \int_0^{2\pi} d\theta \, \frac{\Psi(q, \theta)}{|\mathbf{p} - \mathbf{q}|} \qquad (4.6)$$

in plane polar coordinates. Now introduce a third dimension, giving Cartesian coordinates (p_x, p_y, p_3) and spherical polar coordinates (p', α, ϕ). Project the points of the $p_x p_y$ plane from the point $S(p_x = p_y = 0, p_3 = -p_0)$ onto a sphere of radius p_0. Figure 1 shows a half-plane with ϕ constant (determined by the direction of $\mathbf{p}$). Evidently

$$\mathbf{p} = \lambda\mathbf{OP} + (1 - \lambda)\mathbf{OS} \qquad (4.7)$$

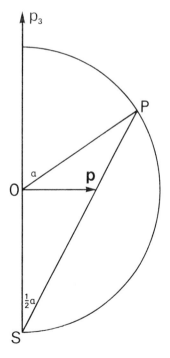

Figure 1.   Projection from plane onto sphere.

for some parameter λ. But as $\mathbf{p} \cdot \mathbf{OS} = 0$,

$$\lambda \mathbf{OP} \cdot \mathbf{OS} = (\lambda - 1)p_0{}^2 \tag{4.8}$$

and so $p^2 = \mathbf{p} \cdot \mathbf{p} = p_0{}^2(2\lambda - 1)$ determines λ.

If $\mathbf{q}$ projects to Q on the sphere and v is the corresponding parameter, then $\mathbf{p} - \mathbf{q} = \lambda \mathbf{OP} - v\mathbf{OQ} + (v - \lambda)\mathbf{OS}$; evaluating $(\mathbf{p} - \mathbf{q}) \cdot (\mathbf{p} - \mathbf{q})$ using (4.8) gives $|\mathbf{p} - \mathbf{q}|^2 = 2\lambda v(p_0{}^2 - \mathbf{OP} \cdot \mathbf{OQ}) = \lambda v|\mathbf{OP} - \mathbf{OQ}|^2$. Finally, putting in the values of λ and v gives

$$4p_0{}^4|\mathbf{p} - \mathbf{q}|^2 = (p^2 + p_0{}^2)(q^2 + p_0{}^2)|\mathbf{OP} - \mathbf{OQ}|^2 \tag{4.9}$$

The transformation is essentially from p to α by $p = p_0 \tan \frac{1}{2}\alpha$, so that

$$dp = \frac{1}{2} p_0 \frac{p^2 + p_0{}^2}{p_0{}^2} \, d\alpha, \; p \, dp = \frac{1}{4} p_0{}^{-2}(p^2 + p_0{}^2)^2 \sin \alpha \, d\alpha.$$

Thus (4.6) can be written in the form (β and θ are the polar angles of $\mathbf{OQ}$)

$$(p^2 - 2\mu E)\Psi(\mathbf{p}) = (p^2 + p_0{}^2)^{-1/2} \frac{P}{2\pi} \int_0^\pi \sin \beta \, d\beta \int_0^{2\pi} d\theta \frac{(q^2 + p_0{}^2)^{3/2}\Psi(\mathbf{q})}{|\mathbf{OP} - \mathbf{OQ}|}$$

which suggests choosing $p_0{}^2 = -2\mu E$. Then

$$\Phi(\alpha, \phi) = \frac{P}{4\pi p_0} \int_0^\pi \sin \beta \, d\beta \int_0^{2\pi} d\theta \frac{\Phi(\beta, \theta)}{\sin \frac{1}{2}\gamma} \tag{4.10}$$

with $\Phi = (p^2 + p_0{}^2)^{3/2}\Psi$, and γ the angle between $\mathbf{OP}$ and $\mathbf{OQ}$. An alternative method (BI66) is to scale the plane coordinates by a factor p_0 and then project onto a unit sphere. Since γ is invariant under three-dimensional orthogonal transformations, equation (4.10) demonstrates the hidden symmetry, the invariance group being O_3.

Equation (4.10) can be solved by expanding the kernel in spherical harmonics. For $\gamma > 0$

$$\operatorname{cosec} \frac{1}{2}\gamma = \sum_{l=0}^\infty 2P_l(\cos \gamma) = 8\pi \sum_{l=0}^\infty (2l + 1)^{-1} \sum_{m=-l}^l Y_l{}^m(\alpha, \phi)\overline{Y}_l{}^m(\beta, \theta)$$

on using the addition theorem (1.28). Thus (4.10) becomes

$$\Phi(\alpha, \phi) = \frac{2P}{p_0} \sum_{ml} \frac{Y_l{}^m(\alpha, \phi)}{(2l + 1)} \int_0^\pi \int_0^{2\pi} \sin \beta \, d\beta \, d\theta \overline{Y}_l{}^m(\beta, \theta)\Phi(\beta, \theta) \tag{4.11}$$

Since the integral is the coefficient c_{lm} of the expansion ($\Phi = \sum c_{lm} Y_l{}^m$) of Φ in spherical harmonics, equating coefficients of $Y_l{}^m(\alpha, \phi)$ gives

$$\frac{2P}{p_0(2l + 1)} c_{lm} = c_{lm}$$

Either $c_{lm} = 0$ or $c_{lm} \neq 0$ and $p_0 = 2P/(2l + 1)$, that is, $E = -2P^2/\mu(2l + 1)^2$ are the energy eigenvalues ($l = 0, 1, 2, \ldots$). With this value of p_0, $Y_l^m(\alpha, \phi)$ is evidently a solution of (4.11) for each m.

The transformation of a scalar product of states of the same energy is

$$\int_0^{2\pi} \int_0^{\pi} \bar{\Phi}_1 \Phi_2 \sin \alpha \, d\alpha \, d\phi = 4p_0^2 \int_0^{2\pi} \int_0^{\infty} (p^2 + p_0^2) \bar{\Psi}_1 \Psi_2 \, p \, dp \, d\phi$$

The virial theorem deduced from (3.11) is true in two dimensions, so that

$$\int_0^{2\pi} \int_0^{\infty} \bar{\Psi}_1 p^2 \Psi_2 \, p \, dp \, d\phi = p_0^2 \int_0^{2\pi} \int_0^{\infty} \bar{\Psi}_1 \Psi_2 \, p \, dp \, d\phi$$

Renormalizing the definition of Φ to $\Phi = 2^{-3/2} p_0^{-2} (p^2 + p_0^2)^{3/2} \Psi$, which does not alter (4.10), makes the scalar product transformation simply

$$\int_0^{2\pi} \int_0^{\pi} \bar{\Phi}_1 \Phi_2 \sin \alpha \, d\alpha \, d\phi = \int_0^{2\pi} \int_0^{\infty} \bar{\Psi}_1 \Psi_2 \, p \, dp \, d\phi \qquad (4.12)$$

Hence the $2l + 1$ orthonormal spherical harmonics $Y_l^m(\alpha, \phi)$ correspond to $2l + 1$ orthonormal momentum space wave functions Ψ.

The Lie algebra of the rotations in the three-dimensional space will now be identified with the o_3 algebra used in Section 3.2. The rotation group has generators

$$L_x = i \left(\cos \phi \cot \alpha \frac{\partial}{\partial \phi} + \sin \phi \frac{\partial}{\partial \alpha} \right)$$

$$L_y = i \left(\sin \phi \cot \alpha \frac{\partial}{\partial \phi} - \cos \phi \frac{\partial}{\partial \alpha} \right), \qquad L_3 = -i \frac{\partial}{\partial \phi}$$

Putting $\cot \alpha = (p_0^2 - p^2)/2pp_0$ and $2p_0 \, \partial/\partial \alpha = (p^2 + p_0^2) \partial/\partial p$ expresses these operators in terms of the polar coordinates (p, ϕ) of the physical two-dimensional space. After changing to Cartesian coordinates (p_x, p_y), the differential operators can be written as position components. Thus

$$-iL_x = \frac{(p_0^2 - p^2)}{2pp_0} \cos \phi \frac{\partial}{\partial \phi} + \frac{(p_0^2 + p^2)}{2p_0} \sin \phi \frac{\partial}{\partial p}$$

$$-ip_0 L_x = \frac{p_0^2 - p^2}{2p^2} p_x \left(p_x \frac{\partial}{\partial p_y} - p_y \frac{\partial}{\partial p_x} \right) + \frac{p_0^2 + p^2}{2p^2} p_y \left(p_x \frac{\partial}{\partial p_x} + p_y \frac{\partial}{\partial p_y} \right)$$

$$hp_0 L_x = p_x p_y x + \tfrac{1}{2}(p_y^2 - p_x^2) y + \tfrac{1}{2} y p_0^2$$

As in Section 3.2, $p_0^2 = -2\mu E$ can be replaced by $-2\mu H = -p_x^2 - p_y^2 + 2hP(x^2 + y^2)^{-1/2}$. Using $[y, p_y^2] = 2ihp_y$ gives

$$p_0 L_x = h^{-1} p_x(x p_y - y p_x) - i p_y + y P(x^2 + y^2)^{-1/2}$$
$$= \tfrac{1}{2} p_x L + \tfrac{1}{2} L p_x - \tfrac{3}{2} i p_y + y P(x^2 + y^2)^{-1/2}$$

Now L_x operates on the functions Φ, so that the corresponding operator on the functions $\Psi = 2^{3/2} p_0{}^2 (p^2 + p_0{}^2)^{-3/2} \Phi$ is $(p^2 + p_0{}^2)^{-3/2} L_x (p^2 + p_0{}^2)^{3/2}$, and it is necessary to evaluate $(p^2 + p_0{}^2)^{-3/2} y (x^2 + y^2)^{-1/2} (p^2 + p_0{}^2)^{3/2}$. This is

$$\frac{y}{(x^2 + y^2)^{1/2}} + \frac{3 i p_y}{2P}$$

by first using

$$[y, (p^2 + p_0{}^2)^{-3/2}] = ih \frac{\partial}{\partial p_y} (p^2 + p_0{}^2)^{-3/2} = -3 i h p_y (p^2 + p_0{}^2)^{-5/2}$$

and then the fact that $(p^2 + p_0{}^2)$ and $2hP(x^2 + y^2)^{-1/2}$ are interchangeable as the right factor of the expression. Hence the operator in the two-dimensional space corresponding to L_x is J_2 of (3.2), and similarly $L_y \to -J_1$ and $L_3 \to J_3$ ($\mu\omega = p_0$).

4.3. THE THREE-DIMENSIONAL CASE

Equation (4.4) may now be solved by following the treatment of (4.5) in the preceding section. Change to spherical polar coordinates (p, θ, ϕ) and introduce a fourth dimension with Cartesian coordinates (p_x, p_y, p_z, p_4) and hyperspherical polar coordinates $(p', \alpha, \theta, \phi)$. The latter are defined by $p_4 = p' \cos \alpha$, $p = p' \sin \alpha$. The points of the (p_x, p_y, p_z) "plane" are projected from $S(p_x = p_y = p_z = 0, p_4 = -p_0)$ onto a four dimensional sphere of radius p_0. Figure 1 again illustrates this, now showing a half-plane with θ and ϕ constant. The derivation of (4.9) is unchanged, but the volume element now includes $p^2 dp = \frac{1}{8} p_0{}^{-3} (p^2 + p_0{}^2)^3 \sin^2 \alpha \, d\alpha$. Choosing $p_0{}^2 = -2\mu E$, the equation for

$$\Phi = \frac{1}{4} p_0{}^{-5/2} (p^2 + p_0{}^2)^2 \Psi \tag{4.13}$$

becomes

$$\Phi(\alpha, \theta, \phi) = \frac{P p_0}{2\pi^2} \int_0^\pi \sin^2 \alpha' \, d\alpha' \int_0^\pi \sin \theta' \, d\theta' \int_0^{2\pi} d\phi' \frac{\Phi(\alpha', \theta', \phi')}{|\mathbf{OP} - \mathbf{OQ}|^2}$$

$$\tag{4.14}$$

where $(\alpha', \theta', \phi')$ are the polar angles of $\mathbf{OQ}$. This demonstrates the four-dimensional rotational symmetry. The definition of Φ includes the normalization factor required for the scalar products to be invariant as in (4.12).

The solution in terms of four-dimensional spherical harmonics may be obtained by expanding $|\mathbf{OP} - \mathbf{OQ}|^{-2}$ in Gegenbauer polynomials and using the addition theorem (1.29). With γ again the angle between $\mathbf{OP}$ and $\mathbf{OQ}$,

$$|\mathbf{OP} - \mathbf{OQ}|^{-2} = p_0^{-2}(2 - 2\cos\gamma)^{-1} = p_0^{-2}\sum_{n=1}^{\infty} C_{n-1}^{1}(\cos\gamma)$$

$$= 2\pi^2 p_0^{-2} \sum_{nlm} n^{-1} Y_{nlm}(\alpha, \theta, \phi)\overline{Y}_{nlm}(\alpha', \theta', \phi')$$

Substituting into (4.14) gives the equivalent of (4.11), and the same argument shows that $\Phi = Y_{nlm}$ is a solution if $P = np_0$ or $E = -P^2/2\mu n^2$. The degeneracy n^2 is just the number of hyperspherical harmonics of given n.

The operators given in (C.2) of Appendix C are the generators of rotations in the four-dimensional space. By following the method described at the end of Section 4.2, the corresponding operators in three-dimensional space can be found and compared with (3.3). The first steps give $p_0\mathbf{M} = -\mathbf{A} - 2i\mathbf{p}$, and using

$$(p^2 + p_0^2)^{-2} Prr^{-1}(p^2 + p_0^2)^2 = Prr^{-1} + 2i\mathbf{p}$$

shows that $p_0\mathbf{M}$ corresponds to $-\mathbf{A}$. Then $\mathbf{L}$, $\mathbf{M}$, $\mathbf{K}$, and $\mathbf{N}$, defined in Section 1.2, correspond respectively to $\mathbf{L}$, $-2\mathbf{A}_n$, $\mathbf{F}$, and $\mathbf{G}$ of Chapter 3. The hyperspherical harmonics Y_{nlm}, which are eigenfunctions of L^2 and L_z, correspond to the states $|nlm\rangle$. In the solution of (4.14), the Y_{nlm} can be replaced by the Z_{nfg}, which are eigenfunctions of $K_z \leftrightarrow F_z$ and $N_z \leftrightarrow G_z$ and therefore correspond to the $|nfg\rangle$ of Chapter 3.

The parity operator Π corresponds to the hyperspace transformation $p_x \to -p_x$, $p_y \to -p_y$, $p_z \to -p_z$, $p_4 \to p_4$. To include Π in the operators of the invariance group one must take O_4 rather than SO_4.

4.4. THE HYLLERAAS METHOD

In operator form the Schrödinger equation is $\frac{1}{2}p^2\Psi - hPr^{-1}\Psi = \mu E\Psi$. The following steps are designed to make the operator depend on r only through r^2, which is represented by $-h^2\nabla^2$ in momentum space. Applying the operator $2r$ to each side yields $r(p^2 - 2\mu E)\Psi = 2hP\Psi$; then applying the operator $r(p^2 - 2\mu E)$ gives $r(p^2 - 2\mu E)r(p^2 - 2\mu E)\Psi = 4h^2 P^2\Psi$. On using the commutator $[r, p^2] = 2ih(\mathbf{p} \cdot \mathbf{r})r^{-1} - 2h^2 r^{-1}$ the operator on the left side becomes

$$r^2(p^2 - 2\mu E)^2 - 2ih(\mathbf{p} \cdot \mathbf{r})(p^2 - 2\mu E) + 4h^2(p^2 - 2\mu E) \qquad (4.15)$$

and the equation is a second-order linear differential equation.

Because of the obvious spherical symmetry it is natural to solve this in spherical polar coordinates. Alternatively, a geometrical argument (K66) based on the four-dimensional symmetry implies that the equation also separates in toroidal coordinates, provided that it is written as an equation for $(p^2 - 2\mu E)^2\Psi$, which gives a hyperspherical harmonic when projected into the hyperspherical coordinates. The operator (4.15) must then be multiplied from the left by $(p^2 - 2\mu E)^2$ and the result manipulated into the alternative form

$$[(p^2 - 2\mu E)^2 r^2 - 2ih(p^2 - 2\mu E)(\mathbf{p} \cdot \mathbf{r}) - 8\mu h^2 E](p^2 - 2\mu E)^2$$

by using (A.1) of Appendix A. The Schrödinger equation may now be written

$$(p^2 + p_0^2)^2 \nabla^2 V - 2(p^2 + p_0^2)\mathbf{p} \cdot \nabla V + 4(P^2 - p_0^2)V = 0 \qquad (4.16)$$

where $V = (p^2 + p_0^2)^2 \Psi$ and $p_0^2 = -2\mu E \ (p_0 > 0)$.

If (p, θ, ϕ) are spherical polar coordinates, the required toroidal coordinates (ξ, η, ϕ) are given by (M43-54)

$$p \cos \theta = (1 + \sin \xi \sin \eta)^{-1} p_0 \sin \xi \cos \eta$$

$$p \sin \theta = (1 + \sin \xi \sin \eta)^{-1} p_0 \cos \xi \qquad (-\tfrac{1}{2}\pi < \xi \leqslant \tfrac{1}{2}\pi)$$

Wave functions must have period 2π in both η and ϕ. Since

$$\mathbf{p} \cdot \nabla = p \frac{\partial}{\partial p} = -\cos \eta \ \mathrm{cosec} \ \xi \frac{\partial}{\partial \eta} - \sin \eta \ \cos \xi \frac{\partial}{\partial \xi}$$

the Schrodinger equation (4.16) becomes

$$\frac{\partial^2 V}{\partial \xi^2} + \sec^2 \xi \frac{\partial^2 V}{\partial \phi^2} + \mathrm{cosec}^2 \ \xi \frac{\partial^2 V}{\partial \eta^2} + (\cot \xi - \tan \xi)\frac{\partial V}{\partial \xi} = \left(1 - \frac{P^2}{p_0^2}\right)V$$

which is easily shown to be separable. However, the details of this are not required here to obtain the solutions, because the operator on the left side is just the o_4 Casimir operator as given in (C.6) of Appendix C. Since $V = (p^2 + p_0^2)^2 \Psi$, if (4.16) is projected onto a four-dimensional sphere as in the preceding section, its solutions are hyperspherical harmonics, and so (4.16) is expected to be an eigenvalue equation of an o_4 Casimir operator. The toroidal coordinates used here have been chosen to correspond exactly with the angular coordinates of the system (C.4) describing the four-dimensional space. Since the eigenvalue of the o_4 Casimir operator $-4F^2 = -L^2 - M^2$ is $1 - n^2$, the Coulomb eigenvalues are given by $p_0 = P/n$ as before. In these coordinates, it is natural to take the Z_{nfg} of (1.17) as the n^2 orthogonal eigenfunctions of (4.16).

Multiplying by $(p^2 + p_0^2)^{-2} = \tfrac{1}{4} p_0^{-4}(1 + \sin \xi \sin \eta)^2$ gives orthogonal eigenfunctions in three-dimensional momentum space, and as in the preceding section the additional factor $4p_0^{5/2}$ normalizes them. This gives (B166)

$$\psi_{nfg}(\mathbf{p}) = (-)^{F+g}(2\pi p_0^2)^{-1}(2P)^{1/2}(1 + \sin \xi \sin \eta)^2$$
$$\times d_{gf}^F(2\xi) \exp \{i(f + g)\phi + i(f - g)\eta\} \qquad (4.17)$$

Since $\mathbf{F}$ and $\mathbf{G}$ correspond to $\mathbf{K}$ and $\mathbf{N}$, these $\psi_{nfg}(\mathbf{p})$ are eigenfunctions of F_z and G_z belonging to the eigenvalues f and g, and $F_\pm$, $G_\pm$ are raising and lowering operators. The wave function (4.17) is the momentum space realization of the state $|nfg\rangle$ of Chapter 3.

The geometrical argument (K66) leading to the toroidal coordinates can be illustrated in two dimensions. A solution $u(\xi)\,v(\eta)$ will be zero only if $u(\xi) = 0$ or $v(\eta) = 0$, and so zero only on curves with equations of the form $\xi = k$ or $\eta = k$. Hence any curve on which some solution is zero indicates a suitable coordinate system for separating the equation. The spherical harmonic solutions of (4.10) are zero on certain horizontal circles $\alpha = \kappa$ on the sphere. Their projections on the plane, on which the $\Psi(\mathbf{p})$ are zero, are circles of center O (and radius $k = p_0 \tan \frac{1}{2}\kappa$). This suggests that polar coordinates will separate (4.4). Because of the rotational symmetry, the solutions of (4.10) can equally well be chosen to be spherical harmonics relative to a rotated polar axis. These have zeros on circles in parallel planes perpendicular to the new polar axis. The projections of such circles on the plane are a set of coaxial circles with centers on a line through O. This suggests that bipolar coordinates (M43-54) will separate any equation for $(p^2 + p_0^2)^{3/2}\Psi$.

Alternatively, any four-dimensional coordinate system (E34) which separates the eigenvalue equation of the o_4 Casimir operator gives a coordinate system of the physical space which separates (4.16). Thus the toroidal coordinates may be obtained from (C.4).

Another simple symmetry property of (4.16) appears from the O_4 transformation $p_4 \to -p_4$. In Figure 1, α changes to $\pi - \alpha$, so that the corresponding transformation in the physical space is $\mathbf{p} \to (p_0/p)^2\mathbf{p}$, a geometrical inversion (C16) with respect to a sphere of radius p_0. In spherical polar coordinates, solutions of (4.16) are therefore unchanged (except for phase) by the substitution $p \to p_0^2/p = P^2/n^2 p$. This can be seen in equation (5.6) in the next section.

5. Special Coordinate Systems

5.1. STANDARD WAVE FUNCTIONS

In Chapter 3 the energy eigenvalue problem was solved by an algebraic method based on angular momentum theory. The energy eigenstates belonging to a given eigenvalue form a space $\mathscr{S}$. The most natural bases for this space were discussed in Section 3.4, and denoted by $|nfg\rangle$ and $|nlm\rangle$. The object of this chapter is to give realizations of these states, both by ordinary wave functions of the position coordinates and by momentum space wave functions.

The results obtained below can be regarded as giving the relation between Chapter 3 and the alternative well-known methods using wave functions. With this point of view it seems convenient to take the wave functions used by Bethe and Salpeter (B57) as given standard forms. Note that the spherical harmonics Y_l^m defined in (1.13) differ from Bethe and Salpeter's Y_{lm} in phase : $Y_l^m = (-)^m Y_{lm}$. To reduce (L66D) confusion the relevant formulas will be quoted in terms of the Y_l^m.

A normalized simultaneous eigenfunction of H, L^2, and L_z is

$$u_{nlm}(\mathbf{r}) = (-)^m R_{nl}(r) Y_l^m(\theta, \phi) \tag{5.1}$$

with

$$R_{nl}(r) = \frac{1}{(2l+1)!} \left[\frac{(n+l)!}{(n-l-1)! \, 2n} \right]^{1/2} (2\alpha)^{3/2} e^{-\alpha r} (2\alpha r)^l F(l+1-n, 2l+2; 2\alpha r) \tag{5.2}$$

in which $\alpha = P/nh$, F is a confluent hypergeometric function, and (r, θ, ϕ) are spherical polar coordinates. The factor $(-)^m$ is required by (1.13) if u_{nlm} is to agree exactly with Bethe and Salpeter. The Laguerre polynomials have been avoided because of a difference in notation between quantum-mechanical applications (B57) and most mathematical work (H65).

Another set of eigenfunctions of H is obtained by separating the Schrödinger equation in the parabolic coordinates $\xi = r(1 + \cos\theta)$, $\eta = r(1 - \cos\theta)$, ϕ. The normalized wave functions are (B57)

$$u_{n_1 n_2 m}(\mathbf{r}) = \frac{\alpha^{|m|+3/2}}{(|m|!)^2} \left[\frac{(n_1 + |m|)! \, (n_2 + |m|)!}{n\pi(n_1)! \, (n_2)!} \right]^{1/2} \exp\left(-\frac{1}{2}\alpha\xi - \frac{1}{2}\alpha\eta + im\phi \right)$$
$$\times (\xi\eta)^{\frac{1}{2}|m|} F(-n_1, |m|+1; \alpha\xi) F(-n_2, |m|+1; \alpha\eta) \tag{5.3}$$

Appendix E gives ∇ in these coordinates. The Runge-Lenz vector (3.3) becomes

$$hA_z = \frac{1}{2}\mu(\xi - \eta)H + h^2\left[\frac{\partial}{\partial\xi}\left(\xi\frac{\partial}{\partial\xi}\right) - \frac{\partial}{\partial\eta}\left(\eta\frac{\partial}{\partial\eta}\right) + \frac{1}{4}\left(\frac{1}{\xi} - \frac{1}{\eta}\right)\frac{\partial^2}{\partial\phi^2}\right]$$

Separating variables by writing $u(\xi, \eta, \phi) = u_1(\xi)u_2(\eta)e^{im\phi}$ leads to the equations (B57, S68)

$$h^2\frac{d}{d\xi}\left(\xi\frac{du_1}{d\xi}\right) + \frac{1}{2}\mu E\xi u_1 + \gamma u_1 - \frac{m^2h^2}{4\xi}u_1 = 0$$

$$h^2\frac{d}{d\eta}\left(\eta\frac{du_2}{d\eta}\right) + \frac{1}{2}\mu E\eta u_2 + hPu_2 - \gamma u_2 - \frac{m^2h^2}{4\eta}u_2 = 0$$

(5.4)

in which γ, the separation constant, is $\frac{1}{2}\alpha h^2(2n_1 + |m| + 1)$. In each equation u can be substituted for u_i if the derivatives are made partial, and then $-m^2$, $E\xi$, and $E\eta$ can be replaced by $\partial^2/\partial\phi^2$, ξH, and ηH respectively. Subtracting then shows that (Sc65) $u_{n_1n_2m}$ is an eigenfunction of hA_z belonging to the eigenvalue

$$hP - 2\gamma = \alpha h^2(n - 2n_1 - |m| - 1) = \alpha h^2(n_2 - n_1)$$

The dependence on ϕ is the same as that of u_{nlm}, so that $u_{n_1n_2 m}$ is also an eigenfunction of L_z belonging to the eigenvalue m.

The momentum space wave functions which are simultaneous eigenfunctions of H, L^2, and L_z are

$$\psi_{nlm}(\mathbf{p}) = (-)^m F_{nl}(p) Y_l^m(\theta, \phi) \tag{5.5}$$

with

$$F_{nl}(p) = \left[\frac{2P}{\pi}\frac{(n - l - 1)!}{(n + l)!}\right]^{1/2} 2^{2l+2}l!\ \frac{n^{l+2}p^lP^{l+2}}{(n^2p^2 + P^2)^{l+2}}\ C_{n-l-1}^{l+1}\left(\frac{n^2p^2 - P^2}{n^2p^2 + P^2}\right)$$

(5.6)

where θ, ϕ are now the polar angles of $\mathbf{p}$, and C_{n-l-1}^{l+1} is a Gegenbauer polynomial (H65). Again ψ_{nlm} is exactly the function given by Bethe and Salpeter; when $l = m = 0$, the formulas simplify to

$$\psi_{n00}(\mathbf{p}) = (-)^n\frac{P^2}{i\pi p}\left(\frac{n}{2P}\right)^{1/2}\left[\frac{(P - inp)^{n-1}}{(P + inp)^{n+1}} - \frac{(P + inp)^{n-1}}{(P - inp)^{n+1}}\right] \tag{5.7}$$

The restrictions on the quantum numbers appearing in these three sets of standard wave functions are as follows (all are integers): in (5.1) and (5.5), $|m| \le l < n$; in (5.3) $n = n_1 + n_2 + |m| + 1$, $0 \le n_1 < n - |m|$ and $0 \le n_2 < n - |m|$.

The momentum space wave functions $\psi_{nfg}(\mathbf{p})$ which are simultaneous eigenfunctions of H, L_z, and A_z were derived in the preceding section and given in (4.17) using (1.18). They belong to the eigenvalues $f + g$ of L_z and $p_0(g - f) = P(g - f)/n$ of A_z. Their quantum numbers f, g, and $F = \frac{1}{2}n - \frac{1}{2}$ are either all integers or all half-integers, restricted by $|f| \leq F$ and $|g| \leq F$.

5.2. PHASES FOR FUNCTIONS OF MOMENTUM

Apart from phase factors, the functions $\psi_{nlm}(\mathbf{p})$ defined in (5.5) are a realization of the states $|nlm\rangle$ of Section 3.4, since they are normalized eigenfunctions of H, L^2, and L_z. Similarly the $\psi_{nfg}(\mathbf{p})$ are a realization of the $|nfg\rangle$. In Chapter 4, both sets of functions were related to the four-dimensional spherical harmonics, so that the required phase factors have already been anticipated in (4.17), (1.17), and (1.15). It is therefore convenient to treat momentum space wave functions before functions of position. The method below verifies the phases shown in (1.15) and (1.17).

For the benefit of readers wishing to pass over the derivation of the phase factors, the results are now summarized. The realization of the $|nfg\rangle$ is given by (5.21), (5.16), and (5.3), or in momentum space by (5.14), (4.17), and (1.18). The realization of the $|nlm\rangle$ is given by (5.21), (5.1), (5.2), and (1.13), or in momentum space by (5.14), (5.5), (5.6), and (1.13).

In any realization of the $|nfg\rangle$, phase factors are required in order to satisfy (2.3):

$$F_{\pm}|n\ f\ g\rangle = (F \mp f)^{1/2}(F \pm f + 1)^{1/2}|n\ f \pm 1\ g\rangle = f_{\pm}|n\ f \pm 1\ g\rangle$$

$$G_{\pm}|n\ f\ g\rangle = (F \mp g)^{1/2}(F \pm g + 1)^{1/2}|n\ f\ g \pm 1\rangle = g_{\pm}|n\ f\ g \pm 1\rangle \quad (5.8)$$

defining the positive numbers $f_{\pm}$ and $g_{\pm}$. In the toroidal coordinates used in Section 4.4 the operators are

$$F_{\pm} = \frac{1}{2} e^{\pm i\phi}\left[\frac{2i\cos\xi}{1 + \sin\xi\sin\eta} + e^{\pm i\eta}\left(\mp\frac{\partial}{\partial\xi} - i\cot\xi\frac{\partial}{\partial\eta} - L_z\tan\xi\right)\right]$$

$$\quad (5.9)$$

$$G_{\pm} = \frac{1}{2} e^{\pm i\phi}\left[\frac{-2i\cos\xi}{1 + \sin\xi\sin\eta} + e^{\mp i\eta}\left(\mp\frac{\partial}{\partial\xi} + i\cot\xi\frac{\partial}{\partial\eta} - L_z\tan\xi\right)\right]$$

These formulas can be obtained directly by expressing $\mathbf{L}$ and $\mathbf{A}$ of Section 3.3 in the coordinates. Alternatively (cf. the end of Section 4.3) they are $(p^2 + p_0^2)^{-2}K_{\pm}(p^2 + p_0^2)^2$ and $(p^2 + p_0^2)^{-2}N_{\pm}(p^2 + p_0^2)^2$ with $K_{\pm}$ and $N_{\pm}$ given in (C.5), and $p^2 + p_0^2 = 2p_0^2(1 + \sin\eta\sin\xi)^{-1}$.

To verify the phase in (4.17), it is not necessary to calculate in full the action of the operators (5.9), because their effect is known from (5.8) except for the phase factor. It is sufficient to look at the lowest power of ξ in (4.17). From (5.9), the parts of the operators that reduce the power by 1 are

$$
F_\pm \sim \frac{1}{2} e^{\pm i\phi \pm i\eta} \left(\mp \frac{\partial}{\partial \xi} - i\xi^{-1} \frac{\partial}{\partial \eta} \right), \quad G_\pm \sim \frac{1}{2} e^{\pm i\phi \mp i\eta} \left(\mp \frac{\partial}{\partial \xi} + i\xi^{-1} \frac{\partial}{\partial \eta} \right)
$$

(5.10)

where $\sim$ means form as $\xi \to 0$. These parts need only be applied to the lowest power of ξ in (4.17), and factors not containing ξ, f, or g are irrelevant; hence it is sufficient to consider

$$
\psi_{nfg} \sim (-)^{F+g} \left[\frac{(F+f)!(F-g)!}{(F+g)!(F-f)!} \right]^{1/2} \frac{\xi^{f-g}}{(f-g)!} \exp \{ i(f+g)\phi + i(f-g)\eta \}
$$

(5.11)

in which the factorials come from (1.18). This has assumed $f \geq g$ so that the summation in (1.18) begins with $t = 0$; it begins with $t = g - f$ if $g \geq f$. Now applying (5.10) to (5.11) shows that F_+ and G_- eliminate the lowest power ξ^{f-g} (giving no information); that $F_- \psi_{nfg} = f_- \psi_{n\,f-1\,g}$; and that $G_+ \psi_{nfg} = g_+ \psi_{n\,f\,g+1}$. Without the phase factor, G_+ would give $-g_+ \psi_{n\,f\,g+1}$.

The known normalization factors f_- and g_+ follow from the $[\ ]^{1/2}$ term in (5.11), checking the validity of the method. To consider the phase only, it is enough to put $\partial/\partial\xi = (f-g)\xi^{-1}$ and $\partial/\partial\eta = i(f-g)$ in (5.10). When $f \leq g$, $(-\xi)^{g-f}/(g-f)!$ replaces $\xi^{f-g}/(f-g)!$ in (5.11). Then putting $\partial/\partial\xi = (g-f)\xi^{-1}$ and $\partial/\partial\eta = -i(g-f)$ in (5.10) gives by inspection that F_+ and G_- give no sign change.

It remains to find the phase factor to attach to $\psi_{nlm}(\mathbf{p})$ of (5.5) to give a realization of $|nlm\rangle$. As discussed in Section 3.4, this factor is determined by (4.17) and (3.9) through the phase convention for the Clebsch-Gordan coefficients. The dependence on m is obviously $(-)^m$ in order to satisfy (2.3). The dependence on n and l is determined by (2.6), that is,

$$
\langle F\ f{=}F\ \ F\ \ g{=}L{-}F \mid F\ F\ \ L\ m{=}L \rangle
$$

is real and positive for every L. This Clebsch-Gordan coefficient appears in the expansion

$$
|n\ f{=}F\ \ g{=}L{-}F\rangle = \sum_l |n\ l\ m{=}L\rangle\langle F\ F\ l\ m{=}L|F\ f{=}F\ \ F\ g{=}L{-}F\rangle
$$

(5.12)

so consider the expansion of $\psi_{n\ f=F\ g=L-F}$ in terms of the $\psi_{n\ l\ m=L}$. The only point of interest is the sign of the coefficient of the $l = L$ term. From (5.6),

p^l is the lowest power of p in ψ_{nlm}, so p^L only occurs in the first ($l = L$) term of the sum, and it is sufficient to look at the coefficient of p^L. As

$$C_{n-L-1}^{L+1}(-1) = (-)^{n-L-1}\binom{n+L}{2L+1}$$

so from (5.6) the coefficient of p^L in F_{nL} has the sign $(-)^{n-L-1}$. From (5.5) and (1.13) the coefficient of p^L in $\psi_{n\ L=m}$ also has this sign. On the other hand, as $p \to 0$, $\xi \to \frac{1}{2}\pi$, $\eta \to \frac{1}{2}\pi$, $e^{i\eta} \to i$, and $p_0(\frac{1}{2}\pi - \xi) \sim 2p\sin\theta$. If $f = F$ the sum in (1.18) has only the $t = 0$ term, and if also $g = L - f$ the lowest power of $(\frac{1}{2}\pi - \xi)$ is $(\frac{1}{2}\pi - \xi)^L$. So when $\psi_{n\ f=F\ g=L-F}$ is expressed in spherical polar coordinates the lowest power of p is p^L, and from (4.17) its coefficient is $(-)^L i^{2F-L} C$ with $C > 0$. The signs of the coefficients of p^L show that

$$\psi_{n\ f=F\ g=L-F} = \sum c_l\ \psi_{n\ l\ m=L}$$

with $c_L = i^{1-L-n} C'$ and $C' > 0$. Comparing with (5.12), in which the corresponding coefficient is real and positive, shows that

$$|n\ L\ m = L\rangle = i^{1-L-n}\psi_{n\ L\ m=L} \text{ (for any } L).$$

The dependence on m mentioned above requires this to be written as

$$|nlm\rangle = i^{l+1-n+2m}\psi_{nlm}.$$

Substituting $\Psi = i^{l+1-n+2m}\psi_{nlm}$ in (4.13) leads to an expression for the four-dimensional spherical harmonic Φ. Equation (1.15) appears by putting $p = p_0\tan\frac{1}{2}\alpha$ and (H65)

$$C_{n-l-1}^{l+1}(-\cos\alpha) = (-)^{n-l-1}\ C_{n-l-1}^{l+1}(\cos\alpha) \tag{5.13}$$

The phase factor just found will be simplified by multiplying by i^{n-1}, which is independent of f, g, and l and can be introduced into all states without affecting the action of $F_\pm$, $G_\pm$, and $L_\pm$. The final results are then

$$|nlm\rangle = i^l(-)^m\psi_{nlm}(\mathbf{p}), \qquad |nfg\rangle = i^{n-1}\psi_{nfg}(\mathbf{p}) \tag{5.14}$$

Multiplying by i^{1-n} instead of i^{n-1} would give phases corresponding to those used by Biedenharn (B61) in treating the hyperspherical harmonics.

5.3. PHASES FOR FUNCTIONS OF POSITION

The functions $u_{n_1n_2\,m}(\mathbf{r})$ and $u_{nlm}(\mathbf{r})$ defined in (5.3) and (5.1) can also represent $|nfg\rangle$ and $|nlm\rangle$ when multiplied by suitable phase factors. For consistency, these realizations should give (5.14) when transformed by (4.1). This condition will determine the dependence on n of the phase factors. To evaluate $U_{nlm}(\mathbf{p})$, the transform of (5.1) by (4.1), needs complicated integrations (P29), but the result must be $\psi_{nlm}(\mathbf{p})$ apart from a phase factor, and in the case $l = m = 0$ considering the lowest power of p only shows that the phase factor is $+1$. The realization by functions of position should therefore give

$$|n \; l=m=0\rangle = u_{n \; l=m=0}(\mathbf{r}) \tag{5.15}$$

The method of Section 5.2 may now be followed. The relations between the quantum numbers f, g and n_1, n_2, m are

$$f = \tfrac{1}{2}(n_1 - n_2 + m), \quad g = \tfrac{1}{2}(n_2 - n_1 + m), \quad n = n_1 + n_2 + |m| + 1$$
$$n_1 = \tfrac{1}{2}n - \tfrac{1}{2} - g = F - g, \quad n_2 = F - f \quad (m \geq 0) \tag{5.16}$$
$$n_1 = F + f, \quad n_2 = F + g \quad (m \leq 0)$$

The operators $F_\pm$ and $G_\pm$ were given in parabolic coordinates by Hughes (H67), but are written here in a convenient factorized form:

$$F_\pm = -\frac{1}{4}\alpha^{-1}(\xi\eta)^{-1/2}e^{\pm i\phi}\left(L_z \mp 2\xi\frac{\partial}{\partial\xi} + \alpha\xi\right)\left(L_z \mp 2\eta\frac{\partial}{\partial\eta} - \alpha\eta\right) \tag{5.17}$$

Replacing α by $-\alpha$ gives $G_\pm$. The factorization is essentially a consequence of the algebra o_4 being the direct sum of two o_3 algebras. In parabolic coordinates the Coulomb problem is therefore equivalent to two two-dimensional harmonic oscillators (R67). The details are given in Section 5.5, where the factors appear naturally as the oscillator operators $A_\pm$ mentioned in Section 3.1. Again only lowest powers in (5.3) need be inspected, and since the coefficient of $(\xi\eta)^{\frac{1}{2}|m|}$ is positive, any phase change on applying $F_\pm$ results from (5.17). Thus putting $L_z = m$ and $2\xi(\partial/\partial\xi) = 2\eta(\partial/\partial\eta) = |m|$ in (5.17) gives by inspection the results in which $|m|$ is decreased:

$$F_- \; u_{n_1 n_2 m} = -f_- \; u_{n_1 \, n_2+1 \; m-1} \quad (m > 0)$$
$$G_- \; u_{n_1 n_2 m} = g_- \; u_{n_1+1 \; n_2 \; m-1} \quad (m > 0)$$
$$F_+ \; u_{n_1 n_2 m} = -f_+ \; u_{n_1+1 \; n_2 \; m+1} \quad (m < 0) \tag{5.18}$$
$$G_+ \; u_{n_1 n_2 m} = g_+ \; u_{n_1 \; n_2+1 \; m+1} \quad (m < 0)$$

The effect of $F_+ F_-$, etc., implies the inverse results:

$$F_+ \; u_{n_1 n_2 m} = -f_+ \; u_{n_1 \; n_2-1 \; m+1} \quad (m \geq 0), \ldots \tag{5.19}$$

The $f_\pm$, $g_\pm$ are the positive normalization factors in (5.8).

The unwanted minus signs in (5.18) and (5.19) can be removed by multiplying $u_{n_1 n_2 m}$ by a factor $(-)^{n_2}$ when $m \geq 0$ and by $(-)^{n_1}$ when $m \leq 0$, but to agree at $m = 0$, when $n_2 = n - n_1 - 1$, the factor $(-)^{n_1}$ must be changed to $(-)^{n-n_1-1} = (-)^{n_2+|m|}$. The two cases may be expressed (BK67, H67) by the single formula

$$(-)^{n_2 + \frac{1}{2}(|m|-m)} = (-)^{F-f} \tag{5.20}$$

The phase factors for the $u_{nlm}(\mathbf{r})$ also follow exactly as in Section 5.2. The $(-)^m$ in (5.1) must be removed to get the standard results with $L_\pm$. The dependence on n and l is found by considering the expansion of $u_{n_1 = 2F-L\ n_2 = 0\ m = L}$ as in (5.12), looking only at the lowest powers $(\xi\eta)^{\frac{1}{2}L}$ and r^L. From (5.3) the coefficient of $(\xi\eta)^{\frac{1}{2}L}$ has positive sign, and from (5.1), (5.2), and (1.13) the coefficient of $r^L \sin^L \theta$ in $u_{n\ l=m=L}$ is also positive. Hence $u_{n_1 = n-1-L\ n_2 = 0\ m = L}$ and $u_{n\ l=m=L}$ need the same phase factor, which is $+1$ using (5.20). The dependence on m already mentioned requires this to be written $(-)^{l+m}$. Then (5.15) holds, so that no dependence on n is necessary.

The results are

$$|nlm\rangle = (-)^{l+m}u_{nlm}(\mathbf{r}), \qquad |nfg\rangle = (-)^{F-f}u_{n_1 n_2 m}(\mathbf{r}) \qquad (5.21)$$

with f, g related to n_1, n_2, m by (5.16).

For the $|nlm\rangle$ the realizations given by Barut (BK67b) have a different phase factor. The reason is that the phase convention for the Clebsch-Gordan coefficients depends on the order of addition of the angular momenta, and Barut takes $\mathbf{L} = \mathbf{G} + \mathbf{F}$. By (2.8) this results in a phase difference of $(-)^{2F-l} = (-)^{n-1-l}$. This would also apply to the work of Hughes (H67); however his results are inconsistent (H68) with the standard phase convention of the Clebsch-Gordan coefficients.

5.4. APPLICATIONS

Some results on Clebsch-Gordan coefficients follow from these explicit realizations of the states. The phase of the first term in the sum in (5.12) has already been discussed. If the proper normalization factors are retained during this investigation, an explicit expression for the Clebsch-Gordan coefficient is obtained:

$$\langle Ff=F\ Fg=L-f\,|\,F\ F\ L\ m=L\rangle = \left[\frac{(2F)!\,(2L+1)!}{L!\,(2F+1+L)!}\right]^{1/2} \qquad (5.22)$$

This follows independently either from the functions of position or from the functions of momentum.

Other results were obtained by Stone (S56) from the identity obtained by substituting (5.14) into (3.9). For example, suppose that $\xi = 0$ is taken in the right side and $p = p_0$, $\theta = \pi/2$ in the left side. When $\xi = 0$, the sum in (1.18) is zero unless $t = \frac{1}{2}(g-f)$, but the $t!\,(t-g+f)!$ in the denominator only allows this if $f = g$; and $f = g$ only appears in the sum in (3.9) if $n - m$ is odd. From (1.13), $\sin^m \theta\ Y_l^m(\theta, \phi)$ is an odd function of $\cos \theta$ if $l - m$ is odd, so then $\psi_{nlm} = 0$ when $\theta = \frac{1}{2}\pi$. From (5.13) $C_{n-l-1}^{l+1}(0) = 0$ if $n - l$ is even, so (5.6) then gives $\psi_{nlm} = 0$ when $p = p_0$. If $n - m$ is even, either $n - l$ is even, or $l - m$

is odd, and both sides of the identity are zero. When $n - m$ is odd and $n - l$ is even, the left side is zero, and so $\langle F\ f\ F\ f\,|\,F\ F\ l\ 2f\rangle = 0$ when $2F - l$ is odd, in agreement with (2.8). When $n - m$ and $n - l$ are both odd, (1.14) and (5.6) with $np = P$ lead to an explicit expression for the coefficient:

$$\langle F\ f\ F\ f\,|\,F\ F\ l\ 2f\rangle =$$

$$\frac{(-)^{F-\frac{1}{2}l}(F + \frac{1}{2}l)!}{(F - \frac{1}{2}l)!\,(\frac{1}{2}l + f)!\,(\frac{1}{2}l - f)!}\left[\frac{(2l + 1)(2F - l)!\,(l + 2f)!\,(l - 2f)!}{(2F + l + 1)!}\right]^{1/2}$$

which is real since $2F - l$ is assumed even.

Conversely, angular momentum theory gives some properties of the wave functions. Equation (5.8) implies recurrence relations between confluent hypergeometric functions when applied in position space. The values of some physically significant integrals are considered next.

The method has essentially determined the Fourier transforms of the four sets of wave functions without integration. From (5.14) and (5.21) the transform by (4.1) of $u_{nlm}(\mathbf{r})$ is $i^{-l}\psi_{nlm}(\mathbf{p})$, agreeing with the results of integration (P29, S65a), and the transform by (4.1) of $u_{n_1n_2\,m}(\mathbf{r})$ is $i^{2f}\psi_{nfg}(\mathbf{p})$. The latter has never been calculated by integration.

Tarter (T70) evaluated $A_{nlm}^{n_1n_2} = \int \bar{u}_{n_1n_2\,m}u_{nlm}\,d\mathbf{r}$ by direct integration, although Park (P60) had previously pointed out that the result was a Clebsch-Gordan coefficient. Equation (5.21) gives

$$A_{nlm}^{n_1n_2} = (-)^{l+m+F-f}\langle F\ f\ F\ m - f\,|\,F\ F\ l\ m\rangle$$

with $2F + 1 = n$, $2f = n_1 - n_2 + m$. Tarter's numerical results have thus been checked against tabulated Clebsch-Gordan coefficients, while his algebraic result follows from (2.9). Roberts (R65) used Tarter's result to deduce the phase factors for functions of position without explicitly considering the phase convention (2.6) for the Clebsch-Gordan coefficients.

The matrix elements given in (3.14) and (3.15) can also be given by integrals. Using (5.1) and (5.21)

$$\langle n\ l\ M\,|\,r_m\,|\,n\ l + 1\ M - m\rangle = -\int_0^\infty R_{nl}R_{n\ l+1}r^3\,dr$$

$$\times \int_0^\pi \sin\theta\,d\theta \int_0^{2\pi} d\phi\,\overline{Y}_l^M\,\frac{r_m}{r}\,Y_{l+1}^{M-m}$$

Then (2.12) and (2.14) show that

$$\langle n\ l\|\mathbf{r}\|n\ l + 1\rangle = (l + 1)^{1/2}\int_0^\infty R_{nl}R_{n\ l+1}r^3\,dr$$

Comparing (3.14) with (63.5) of Bethe and Salpeter suggests that their equation has incorrect sign, which is confirmed on calculating special cases of the integral. The results (3.15) do not seem to have been given before, but can presumably be obtained as the limit $n' \to n$ of the expressions (B57) for $\langle n_1 n_2 \, m \,|\, \mathbf{r} \,|\, n_1' n_2' \, m' \rangle$.

5.5. CONNECTION WITH THE TWO-DIMENSIONAL OSCILLATOR

The connection between the two-dimensional harmonic oscillator problem and the two-dimensional Coulomb problem was given in Section 3.2 by comparing their Schrödinger equations in suitably chosen coordinates. The three-dimensional Coulomb problem was related to the two-dimensional harmonic oscillator by Ravndal and Toyoda (R67), whose method is described below, but their coordinates will be changed so as to use the parabolic coordinates of Section 5.1. This requires treating the oscillator in the modified polar coordinates $\xi = u^2 + v^2$, $\tan \phi = v/u$, where (u, v) are Cartesian coordinates for the oscillator as in Section 3.1.

Separating the oscillator Schrödinger equation in these coordinates shows that the energy eigenfunctions have the form $R(\xi)e^{im\phi}$, where m is an integer. These functions are also eigenfunctions of the angular momentum $L = -i(\partial/\partial\phi)$, and are therefore a realization of the eigenstates of J_3 in (3.1). The radial function R satisfies

$$\frac{d}{d\xi}\left(\xi \frac{dR}{d\xi}\right) - \frac{\mu^2 \omega^2}{4h^2}\, \xi R + \frac{\mu E'}{2h^2}\, R - \frac{m^2}{4\xi}\, R = 0 \qquad (5.23)$$

where μ, ω, and E' are respectively the mass, frequency constant, and oscillator energy eigenvalue. From Section 3.1, $E' = n'h\omega$ with n' a positive integer; also for a given $n' = 2j + 1$, the allowed values of $J_3 = \frac{1}{2}L$ are $j, j - 1, \ldots, -j$. The square-integrable solutions of (5.23) will therefore be written $R_{n'm}$ ($m = n' - 1$, $n' - 3, \ldots, 1 - n'$) and do not depend on the sign of m since (5.23) only involves m^2.

Comparing (5.4) with (5.23) shows that $u_1(\xi) = R_{n'm}(\xi)$ with $u_2(\eta) = R_{n''m}(\eta)$ gives a solution of (5.4) provided $\gamma = \frac{1}{2}\mu E'$ and $hP - \gamma = \frac{1}{2}\mu E''$, that is, $\mu E' + \mu E'' = 2Ph$ or $\mu\omega(n' + n'') = 2P$. The Coulomb eigenvalue E is $-\frac{1}{2}\mu\omega^2 = -2P^2/\mu(n' + n'')^2$. The positive integers n' and n'' have the same parity, since both have opposite parity to m. Thus $n' + n''$ is even, and $n = \frac{1}{2}(n' + n'')$ is a positive integer. The Coulomb wave functions with energy $E = -P^2/2\mu n^2$ correspond to those of an oscillator with frequency constant $\omega = P/\mu n$.

The degeneracy of the Coulomb eigenvalue follows from the degeneracy n' of the oscillator eigenvalue $n'h\omega$. For given n the allowed values of (n', n'') are

$$(1, 2n - 1), (2, 2n - 2), \ldots, (n - 1, n + 1), (n, n), (n + 1, n - 1), \ldots, (2n - 1, 1)$$

The number of values of m is $\min(n', n'')$ so that the Coulomb degeneracy is $(2\sum_{n'=1}^{n-1} n') + n = n^2$.

The quantum numbers n' and n'' will now be identified. Since $\gamma = \frac{1}{2}\mu n' h\omega$, the remarks after (5.4) now show that $R_{n'm}(\xi)R_{n''m}(\eta)e^{im\phi}$ is an eigenfunction of A_z belonging to the eigenvalue $P - \mu n'\omega = P(1 - n'/n)$. Then $n - n' = g - f$, and also $n' + n'' = 2n$, so that

$$n' = n - g + f = 2n_1 + |m| + 1, \qquad n'' = n + g - f = 2n_2 + |m| + 1$$
$$(5.24)$$

If $R_{nm}(\xi)e^{im\phi}$ is a normalized oscillator wave function, then $\pi\int_0^\infty |R_{nm}|^2 d\xi = 1$. For the Coulomb problem the volume element is $\frac{1}{4}(\xi + \eta)\, d\xi\, d\eta\, d\phi$, and to consider normalization also requires $\int_0^\infty |R_{nm}|^2 \xi\, d\xi$. This is $nh\omega/\pi\mu\omega^2$, since (e.g., from the virial theorem) the expectation value of the oscillator potential energy $\frac{1}{2}\mu\omega^2\xi$ is half the oscillator energy. The Coulomb normalization integral is therefore

$$\|u\|^2 = 2\pi \int_0^\infty d\xi \int_0^\infty d\eta\, |R_{n'm}(\xi)R_{n''m}(\eta)|^2 \frac{1}{4}(\xi + \eta)$$

$$= \frac{(n' + n'')h}{2\pi\mu\omega} = \frac{hn^2}{P\pi} = \frac{n}{\alpha\pi}$$

using the α of Section 5.1. The normalized Coulomb eigenfunction is therefore $(\alpha\pi/n)^{1/2} R_{n'm}(\xi)R_{n''m}(\eta)e^{im\phi}$. Comparison with (5.3) shows that $R_{n'm}(\xi)$ is

$$\frac{1}{|m|!}\left[\frac{\alpha(\frac{1}{2}n' + \frac{1}{2}|m| - \frac{1}{2})!}{\pi(\frac{1}{2}n' - \frac{1}{2}|m| - \frac{1}{2})!}\right]^{1/2} (\alpha\xi)^{\frac{1}{2}|m|}e^{-\frac{1}{2}\alpha\xi}$$

$$\times F\left(-\frac{1}{2}n' + \frac{1}{2}|m| + \frac{1}{2}, |m| + 1; \alpha\xi\right)$$

which agrees ($\alpha = \mu\omega/h$) with the result found (S44) by solving (5.23).

The shift operators for the oscillator angular momentum and energy were mentioned in Section 3.1. In the modified polar coordinates they are

$$A_\pm = \frac{1}{2}(\alpha\xi)^{-1/2}e^{\mp i\phi}\left(\alpha\xi + 2\xi\frac{\partial}{\partial\xi} \mp i\frac{\partial}{\partial\phi}\right)$$

$$A_\pm^* = \frac{1}{2}(\alpha\xi)^{-1/2}e^{\pm i\phi}\left(\alpha\xi - 2\xi\frac{\partial}{\partial\xi} \mp i\frac{\partial}{\partial\phi}\right)$$
$$(5.25)$$

with * denoting complex conjugation with respect to the oscillator area element $\frac{1}{2}d\xi\, d\phi$. The corresponding operators with ξ changed to η will be written $B_\pm$ and $B_\pm^*$. Expressions for the Coulomb shift operators $F_\pm$ and $G_\pm$ can now be found.

For example, (5.24) shows that F_+ has to raise n', lower n'', and raise m. For the oscillator wave functions A_+^* raises n' and m, and B_- lowers n'' and raises m. Then $B_-(R_{n'm}R_{n''m}e^{im\phi})$ is a multiple of $R_{n'm}R_{n''-1\ m+1}\,e^{i(m+1)\phi}$. Since $R_{n'm}e^{i(m+1)\phi}$ is not an oscillator wave function, one must multiply by $e^{-i\phi}$ before using A_+^*. Thus $F_+ = kA_+^*\,e^{-i\phi}B_-$, where k is a number that can be determined by considering either the term containing $\partial^2/\partial\phi^2$ or the term containing no derivative. The latter is $\frac{1}{4}k\alpha(\xi\eta)^{1/2}e^{i\phi}$in $kA_+^*\,e^{-i\phi}B_-$, and in F_+ comes from $-A_{n+}$, using the definition in Section 3.3. To identify the term in $-A_{n+}$, replace μH by $-P^2/2n^2 (= -\frac{1}{2}\alpha^2h^2)$ in $h\mathbf{A}$ given in (3.3) and multiply by $(-n/2Ph)$. The $+$ component then contains $\frac{1}{4}\alpha(x+iy) = \frac{1}{4}\alpha r \sin\theta\,e^{i\phi} = \frac{1}{4}\alpha(\xi\eta)^{1/2}e^{i\phi}$, so that $k = 1$. Similarly

$$F_+ = A_+^*\,e^{-i\phi}B_- = B_-\,e^{-i\phi}A_+^*\,, F_- = B_-^*\,e^{i\phi}A_+ = A_+\,e^{i\phi}B_-^*$$

$$G_+ = -A_-\,e^{-i\phi}B_+^* = -B_+^*\,e^{-i\phi}A_-\,, G_- = -B_+\,e^{i\phi}A_-^* = -A_-^*\,e^{i\phi}B_+$$
$$(5.26)$$

which become (5.17) on substituting (5.25). Another way of finding k, which will be used in the next chapter for energy lowering operators, is to compare the known effect of F_+ on $u_{n_1 n_2\,m}$ [equations (5.8), (5.16), and (5.21)] with the effect of A_+^* and B_- on R_{nm} (given at the end of Section 6.2).

From Section 3.2, $(\frac{1}{2}\xi, 2\phi)$ correspond to the usual plane polar coordinates (r, θ) for the two-dimensional Coulomb problem; hence the area element is $\frac{1}{2}\xi\,d\xi\,d\phi$. The expectation value $\frac{1}{2}nh\omega$ of the oscillator potential energy now shows that if ψ_n is normalized in the oscillator problem, then $\|\psi_n\|^2 = (\mu\omega^2)^{-1}\frac{1}{2}nh\omega$ in the two-dimensional Coulomb problem. Therefore, $(4P/n^2h)^{\frac{1}{2}}R_{nm}(2r)e^{\frac{1}{2}im\theta}$, with $\omega = 2P/\mu n$ as in Section 3.2, is a normalized Coulomb eigenfunction. This agrees with the direct solution (Z67) in polar coordinates. Since n is odd, m is even and $\frac{1}{2}m$ is an integer. The angular momentum of an oscillator state is twice that of the corresponding Coulomb state.

6. Noninvariance Algebras and Groups

Matrix elements of $\mathbf{r}$ between states of the same energy were obtained in Section 3.6. In this chapter all matrix elements of $\mathbf{r}$, $\mathbf{p}$, $1/r$, and e^{ikz} are calculated, between arbitrary bound states. The results for e^{ikz}, or transition form factors, are given in (6.29) in a form somewhat more symmetric than the original calculations (BK67b). The procedure uses an $o_{4, 2}$ algebra, called the noninvariance algebra of the hydrogen atom, whose complexification includes shift operators between states of different energy. The bound states thus span the domain of an irreducible representation of $o_{4, 2}$.

However, the calculation is not presented here in a purely algebraic way, since I have used the connection between Coulomb and oscillator eigenfunctions, and this was established in Section 5.5 by comparing the Schrödinger equations of the two systems. The states $|nfg\rangle$ are therefore used, but matrix elements relative to the $|nlm\rangle$ may then be obtained using Clebsch-Gordan coefficients.

Following the plan of Chapter 3, the concept of a noninvariance algebra is first illustrated by simpler systems—harmonic oscillators and the two-dimensional Coulomb problem. The only part of these first three sections which is used for the hydrogen atom is the discussion of A_{+} and A_{+}^{*} at the end of Section 6.2.

Because the parameter α in the hydrogenic wave functions depends on n, shift operators between states of different energy contain scaling operators S', which make the matrix elements very complicated. However, the exponential form of S' derived in Section 6.5 shows (BK67a) that it is an element of the group generated by the $o_{4, 2}$ algebra. This allows its matrix elements to be written down from the results in Sections 1.13 or 2.5. A similar technique deals with e^{ikz}, which is also an element of the group.

6.1. THE ONE-DIMENSIONAL HARMONIC OSCILLATOR

If A and A^{*} are the shift operators of a one-dimensional harmonic oscillator, as in Section 3.1, the operators (L66)

$$T_{1} = \tfrac{1}{4}(A^{*2} + A^{2}), \qquad T_{2} = \tfrac{1}{4}i(A^{2} - A^{*2}), \qquad T_{3} = \tfrac{1}{2}A^{*}A + \tfrac{1}{4} \quad (6.1)$$

are a basis of an $o_{2,1}$ algebra. The commutation relations (1.31) are satisfied by $E_{23} = -iT_1$, $E_{31} = iT_2$, and $D_{12} = -iT_3$. This is a noninvariance algebra for the oscillator, since T_1 and T_2 do not commute with the energy. The complexification of the algebra contains the operators $2(T_1 + iT_2) = A^{*2}$ and $2(T_1 - iT_2) = A^2$ which raise and lower the energy by $2\hbar\omega$, and so connect states of the same parity. The parity operator commutes with the algebra; hence the even and odd functions each form an invariant space, giving irreducible representations of class D_k^+ of $o_{2,1}$. The operator T_3 has the eigenvalues $\frac{1}{4}, 1\frac{1}{4}, 2\frac{1}{4}, \ldots$ in the even states and eigenvalues $\frac{3}{4}, 1\frac{3}{4}, 2\frac{3}{4}, \ldots$ in the odd states. The Casimir operator $T_1{}^2 + T_2{}^2 - T_3{}^2$ is the number $\frac{3}{16}$.

Since the noninvariance algebra contains the Hamiltonian $2\hbar\omega T_3$, the noninvariance group obtained as an exponential function on the algebra contains the time-evolution operator $\exp(-2i\omega t T_3)$. This suggests that the motion of a harmonic oscillator can be represented as a rotation in a suitable space. However the noninvariance group is not $SO_{2,1}$ because the eigenvalues of T_3 are not integers.

Other noninvariance algebras are obtained from the operators (H66)

$$J_+ = (l + 1 - A^*A)^{1/2}A^*, \qquad J_- = (l - A^*A)^{1/2}A, \qquad J_z = A^*A - \tfrac{1}{2}l \quad (6.2)$$

which satisfy the angular momentum commutation relations for any value of the constant l. If l is an integer, the lowest $l + 1$ energy eigenstates span an invariant space in which $J_+^* = J_-$. The operators (6.2) give an o_3 algebra when their domain is restricted to this $(l + 1)$-dimensional space. In the rest of the space, spanned by states with energy greater than $(l + \frac{1}{2})\hbar\omega$, $J_+^* = -J_-$ and (6.2) give an $o_{2,1}$ algebra and another irreducible representation of type D_k^+.

A knowledge of a noninvariance algebra may allow an algebraic determination of properties involving states of different energy. For example, the o_3 algebra above permits the use of angular momentum theory to get matrix elements of $J_\pm$, hence of A and A^*, hence of position and momentum.

6.2. THE TWO-DIMENSIONAL HARMONIC OSCILLATOR

Using u and v for the Cartesian coordinates, as in Section 3.1, and A_u and A_v for the lowering operators, define $T_{1u} = \frac{1}{4}(A_u^{*2} + A_u{}^2)$, etc., as in (6.1). The T_{iu} form an $o_{2,1}$ algebra, as do the T_{iv}, and the direct sum of these algebras is an $o_{2,2}$ algebra. This does not have an o_3 subalgebra and so does not include the invariance algebra found in Section 3.1. It does contain $J_1 = T_{3u} - T_{3v}$. On taking the operators $A_u^*A_v$ and $A_u A_v^*$ to get J_2 and J_3, the operators $A_u^*A_v^*$ and $A_u A_v$ are also necessary for all commutators to be included. The resulting noninvariance algebra has dimension 10 and contains o_3 and $o_{2,2}$ subalgebras. The simplest algebra with these properties is $o_{3,2}$ (G59).

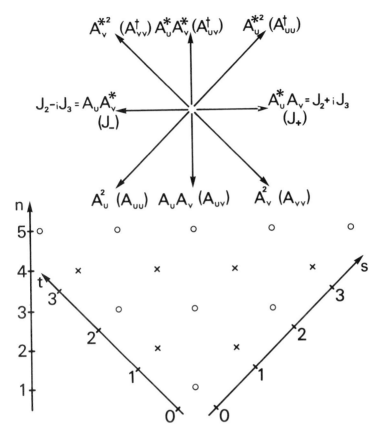

Figure 2. Shift operators and energy eigenstates of the oscillator (first 5 levels) and Coulomb (first 3 levels) problems in two dimensions.

The complexification of the algebra contains the shift operators of the invariance algebra and operators that change the energy by $2\hbar\omega$. As in the one-dimensional case, wave functions of the same parity span an invariant space. The changes of state caused by the shift operators can be indicated by arrows on a diagram in which each point (circles for even parity states, crosses for odd parity states) represents an energy eigenfunction (Figure 2).

The noninvariance algebra is identified as an $o_{3,2}$ algebra by choosing the following Hermitian operators as basis:

$$J_1 = T_{3u} - T_{3v} = \tfrac{1}{2}(A_u^* A_u - A_v^* A_v)$$

$$J_2 = \tfrac{1}{2}(A_u A_v^* + A_u^* A_v)$$

$$J_3 = \tfrac{1}{2}i(A_u A_v^* - A_u^* A_v)$$

$$U_1 = \tfrac{1}{2}i(A_u^* A_v^* - A_u A_v)$$

$$U_2 = T_{2u} - T_{2v} = \tfrac{1}{4}i(A_v^{*2} - A_v^2 - A_u^{*2} + A_u^2)$$

$$U_3 = -T_{1u} - T_{1v} = -\tfrac{1}{4}(A_v^{*2} + A_v^2 + A_u^{*2} + A_u^2) \qquad (6.3)$$

$$V_1 = \tfrac{1}{2}(A_u^* A_v^* + A_u A_v)$$

$$V_2 = -T_{1u} + T_{1v} = \tfrac{1}{4}(A_v^{*2} + A_v^2 - A_u^{*2} - A_u^2)$$

$$V_3 = -T_{2u} - T_{2v} = \tfrac{1}{4}i(A_v^{*2} - A_v^2 + A_u^{*2} - A_u^2)$$

$$S = T_{3u} + T_{3v} = \tfrac{1}{2}(A_v^* A_v + A_u^* A_u + 1)$$

Their commutators agree with Table 1 of Appendix D, with J_α instead of L_α. This algebra can also be exhibited using the shift operators (G68) $A_\pm$ and $A_\pm^*$. The ten possible independent products are a basis for the complexification of the algebra, and the $o_{3,2}$ algebra consists of the subset of Hermitian operators.

Since $A_u^* A_u \, \psi_{st} = s\psi_{st}$, $\|A_u \, \psi_{st}\| = s^{1/2}$ (assuming ψ_{st} normalized) and so the phase of ψ_{st} can be chosen so that $A_u \, \psi_{st} = s^{1/2} \, \psi_{s-1 \, t}$. Similarly $A_u^* \, \psi_{st} = (s+1)^{1/2} \, \psi_{s+1 \, t}$, $A_v \, \psi_{st} = t^{1/2} \, \psi_{s \, t-1}$, $A_v^* \, \psi_{st} = (t+1)^{1/2} \, \psi_{s \, t+1}$, and the effect on ψ_{st} of the ten products $A_u^* A_u$, etc., can be written down. This gives the matrix representation of these operators relative to the states ψ_{st}. When the algebra is described in terms of $A_\pm$ and $A_\pm^*$, it is more convenient to use states ψ_{nM}, which are eigenstates of J_3 belonging to the eigenvalue M. As $A_\pm^* A_\pm \, \psi_{nM} = (S \pm J_3 - \tfrac{1}{2})\psi_{nM} = (\tfrac{1}{2}n - \tfrac{1}{2} \pm M)\psi_{nM}$, the phase of ψ_{nM} can be chosen so that $A_\pm \, \psi_{nM} = (\tfrac{1}{2}n - \tfrac{1}{2} \pm M)^{1/2}\psi_{n-1 \; M\mp 1/2}$ and $A_\pm^* \, \psi_{nM} = (\tfrac{1}{2}n + \tfrac{1}{2} \pm M)^{1/2}\psi_{n+1 \; M\pm 1/2}$. With (L60) the phase factor $(-)^{\frac{1}{2}(n-1)-|M|}$, the functions $R_{n \; 2M}(\xi)e^{2iM\phi}$ of Section 5.5 are a realization of the ψ_{nM}.

The phase factor shows that $A_\pm$ operating on the $R_{nm}(\xi)e^{im\phi}$ produce a sign change if $\tfrac{1}{2}(n-1) - |\tfrac{1}{2}m| - \tfrac{1}{2}(n-2) + |\tfrac{1}{2}m \mp \tfrac{1}{2}|$ is odd. Thus A_+ (hence also A_+^*) gives a sign change if $m \le 0$, and A_- (also A_-^*) gives a sign change if $m \ge 0$. This can be quickly verified from the effect of (5.25) on the lowest power of ξ in $R_{nm}(\xi)$. The action of $A_\pm$ and $A_\pm^*$ is summarized in equation (6.11) and illustrated in Figure 4.

6.3. THE TWO-DIMENSIONAL COULOMB PROBLEM

In Section 3.2 a correspondence was established between the n degenerate Coulomb eigenfunctions belonging to the energy $-2P^2/\mu n^2$ and n eigenfunctions of a harmonic oscillator with frequency constant $\omega = 2P/\mu n$. This correspondence was used to get the generators (3.2) of the Coulomb invariance group from the generators (3.1) of the oscillator invariance group. The non-invariance algebra of the Coulomb problem can be obtained in the same way from the results of the preceding section.

First consider the invariance algebra. The J_i listed in (6.3) are just (3.1). The algebra for the Coulomb case must include extensions of (3.2), so that their domain is the space spanned by all bound state wave functions. This might be accomplished by changing the factor $(2\mu\omega)^{-1}$ to $(-8\mu H)^{-1/2}$, but the operators begin to look complicated, as does an explicit form of the square root of the Hamiltonian (S68a). Here the operators will be defined by stating their action on the normalized Coulomb energy eigenstates $|st\rangle$. For instance J_1 can be defined by $J_1|st\rangle = \frac{1}{2}(s-t)|st\rangle$. To get J_2 and J_3 it is simplest to define the shift operators $J_\pm$ which correspond to the oscillator operators $A_u^* A_v$ and $A_v^* A_u$. Thus $J_+|st\rangle = (st+t)^{1/2}|s+1\ t-1\rangle$ and $J_-|st\rangle = (s+st)^{1/2}|s-1\ t+1\rangle$ are definitions, the factors $s^{1/2}$, etc., reproducing the effects of the oscillator shift operators (M58). These definitions essentially give the matrix representation of the operators relative to energy eigenstates $|st\rangle$, and this representation shows that $J_+^\dagger = J_-$. Then $\frac{1}{2}(J_+ + J_-)$ and $\frac{1}{2}i(J_- - J_+)$ are Hermitian.

The operators that change the energy are also best defined by their effect on $|st\rangle$, which is chosen to agree with that of the corresponding oscillator operators. Using a notation in which A_{uv} corresponds to the oscillator $A_u A_v$, the definitions are

$$A_{uu}|st\rangle = (s^2 - s)^{1/2}|s-2\ t\rangle$$
$$A_{uv}|st\rangle = (st)^{1/2}|s-1\ t-1\rangle \qquad (6.4)$$
$$A_{vv}|st\rangle = (t^2 - t)^{1/2}|s\ t-2\rangle$$

The consequent matrix representations of A_{uu}, A_{uv}, and A_{vv} provide matrix representations of their complex conjugates. For example,

$$\langle s't' | A_{uu}^\dagger | st\rangle = \overline{\langle st | A_{uu} | s't'\rangle} = (s'^2 - s')^{1/2}\delta_{s',s+2}\delta_{t',t}$$

implying

$$A_{uu}^\dagger|st\rangle = (s+2)^{1/2}(s+1)^{1/2}|s+2\ t\rangle$$

and similarly

$$A_{uv}^\dagger|st\rangle = (s+1)^{1/2}(t+1)^{1/2}|s+1\ t+1\rangle$$
$$A_{vv}^\dagger|st\rangle = (t+2)^{1/2}(t+1)^{1/2}|s\ t+2\rangle \qquad (6.5)$$

Finally, define $S|st\rangle = \frac{1}{2}(s+t+1)|st\rangle$. Figure 2 also represents the action of the Coulomb shift operators, which are in brackets, only the circles corresponding to Coulomb eigenstates.

Since the ten operators $J_1, J_+, J_-, A_{uu}, A_{uv}, A_{vv}, A_{uu}^\dagger, A_{uv}^\dagger, A_{vv}^\dagger$, and S have been defined to give the same effect as analogous oscillator operators, they have exactly the same commutators. Thus

$$A_{uu}A_{uv}^{\dagger}|st\rangle = (s+1)^{1/2}(t+1)^{1/2}A_{uu}|s+1\ t+1\rangle = (t+1)^{1/2}s^{1/2}(s+1)|s-1\ t+1\rangle$$

$$A_{uv}^{\dagger}A_{uu}|st\rangle = (s^2-s)^{1/2}A_{uv}^{\dagger}|s-2\ t\rangle = (s-1)s^{1/2}(t+1)^{1/2}|s-1\ t+1\rangle$$

and $[A_{uu},A_{uv}^{\dagger}]|st\rangle = 2s^{1/2}(t+1)^{1/2}|s-1\ t+1\rangle = 2J_-|st\rangle$. This holds for every $|st\rangle$, so $[A_{uu},A_{uv}^{\dagger}] = 2J_-$, corresponding to $[A_u^2, A_u^*A_v^*] = 2J_- = 2A_uA_v^*$ for the oscillator. The Hermitian operators

$$J_1,\ S,\ J_2 = \tfrac{1}{2}(J_+ + J_-),\quad J_3 = \tfrac{1}{2}i(J_- - J_+)$$

$$U_1 = \tfrac{1}{2}i(A_{uv}^{\dagger} - A_{uv}),\quad V_1 = \tfrac{1}{2}(A_{uv}^{\dagger} + A_{uv})$$

$$U_2 = \tfrac{1}{4}i(A_{vv}^{\dagger} - A_{vv} - A_{uu}^{\dagger} + A_{uu}) \tag{6.6}$$

$$U_3 = -\tfrac{1}{4}(A_{vv}^{\dagger} + A_{vv} + A_{uu}^{\dagger} + A_{uu})$$

$$V_2 = \tfrac{1}{4}(A_{vv}^{\dagger} + A_{vv} - A_{uu}^{\dagger} - A_{uu})$$

$$V_3 = \tfrac{1}{4}i(A_{vv}^{\dagger} - A_{vv} + A_{uu}^{\dagger} - A_{uu})$$

have the same commutators as the operators (6.3). This noninvariance algebra of the two-dimensional Coulomb problem is therefore an $o_{3,2}$ algebra.

The group associated with this algebra contains the invariance group SO_3 generated by the J_i. Since $n = s + t + 1$ is odd, the eigenvalues of the compact generator S are half-integers, so that S does not generate an SO_2 subgroup, suggesting that the group is not $SO_{3,2}$.

If the operators V_1, V_2, V_3, and S are removed from (6.6), the remaining J_i and U_i form an $o_{3,1}$ algebra, under which no subspace of the bound state wave functions is invariant. In other words, the representation of $o_{3,2}$ contains an irreducible representation of this $o_{3,1}$ subalgebra. For this reason the $o_{3,1}$ algebra can be taken as the noninvariance algebra, and then the noninvariance group is $SO_{3,1}$ because the compact generators of the $o_{2,1}$ subalgebras are the J_i, which have integer eigenvalues.

As discussed above, the operators J_i of the invariance algebra, extensions of (3.2), can be written as complicated functions of the observables x, p_x, y, and p_y. Doing this for the other operators of the noninvariance algebra is hindered by the normalization of the Coulomb and oscillator wave functions being different and by the frequency of the related oscillator depending on the Coulomb energy. From Section 5.5 the normalized Coulomb eigenfunction is $|st\rangle = (4P/n^2h)^{1/2}\psi_{st}$. To change ψ_{st} to an eigenfunction of an oscillator of a different frequency, a scaling operator $S(\beta)$ is required, defined by $S(\beta)\psi(u, v) = \beta\psi(\beta u, \beta v)$. The factor β makes $S(\beta)$ unitary in the oscillator problem. The ψ_{st} depend on the frequency constant ω by being functions of the dimensionless variable $u(\mu\omega/h)^{1/2}$. This means that if $\psi_{st}(\omega)$ are the eigenfunctions for the oscillator of frequency ω, then $S(\beta)\psi_{st}(\omega)$ are the normalized eigenfunctions for an oscillator of frequency ω' if $\beta = (\omega'/\omega)^{1/2}$. Remembering that ω depends on the Coulomb energy through $\omega = 2P/\mu n$, the energy raising and lowering operators are seen to need $S(\beta)$ with $\beta = [n/(n \pm 2)]^{1/2}$. Section 6.5 will describe the properties of scaling operators, which have also been called scale operators (M66, BI66), dilation operators (P66), and dilatation operators (BK67).

The shift operations $|st\rangle \to f|s't'\rangle$ of the algebra, where f is the required factor appearing in (6.4) or (6.5), can thus be achieved in three stages. First apply the corresponding oscillator shift operator, to get $(4P/n^2h)^{1/2} f \psi_{s't'}$. Next apply the oscillator scaling operator, to get $(4P/n^2h)^{1/2} f \psi_{s't'}$, where $\psi_{s't'}$ is now a normalized oscillator wave function with the required frequency constant $\omega' = 2P/\mu(n \pm 2)$. Finally, multiply by $n/n \pm 2$, to get $f|s't'\rangle$, where $|s't'\rangle$ is the normalized Coulomb eigenfunction. The last two steps may be combined by defining operators $S_{n\pm}$ which multiply by $[n/(n \pm 2)]^{3/2} = \beta^3$ and replace u and v by βu and βv. Using the known expressions for the oscillator shift operators then gives expressions such as

$$A_{uu} = S_{n-}\left(\frac{h}{4}\frac{\partial^2}{\partial u^2}\frac{n}{P} + u\frac{\partial}{\partial u} + \frac{1}{2} + \frac{u^2}{h}\frac{P}{n}\right)$$

in which $P/n = (-\tfrac{1}{2}\mu H)^{1/2}$.

When working with eigenstates $|nM\rangle$ of energy and angular momentum Mh it is preferable to take as the basis of the algebra the operators $A_{\pm}$, etc., corresponding to $A_{\pm}^2$, etc., for the oscillator. The effect of these operators on the $|nM\rangle$ can be written down from the results in the preceding section. However, this has already been done in Section 1.10. One realization of the $|nM\rangle$, obtained in Section 4.2 using the projection from momentum space onto a sphere, consists of the spherical harmonics Y_l^M with $l = \tfrac{1}{2}n - \tfrac{1}{2}$. Then, for example, corresponding to

$$A_-A_+ \, \psi_{nM} = A_-(\tfrac{1}{2}n - \tfrac{1}{2} + M)^{1/2}\psi_{n-1 \, M-1/2}$$
$$= (\tfrac{1}{2}n - M - \tfrac{1}{2})^{1/2}(\tfrac{1}{2}n - \tfrac{1}{2} + M)^{1/2}\psi_{n-2 \, M}$$

for the oscillator, there is $A_0 Y_l^M = (l^2 - M^2)^{1/2} Y_{l-1}^M$, the first of equations (1.36). Equations (1.38) are the analogs of (6.6).

6.4. THE NONINVARIANCE ALGEBRA OF THE THREE-DIMENSIONAL PROBLEM

In Section 6.3 the operators spanning the noninvariance algebra were defined by analogy with the corresponding oscillator operators. The connection established in Section 5.5 between the three-dimensional Coulomb problem and the two-dimensional oscillator may be used in the same way to suggest the operators required for the noninvariance algebra of the Coulomb problem. In this section the definitions of these operators will be written down, then the algebra will be identified, and finally the correspondence with the oscillator will be discussed.

It is the Coulomb eigenfunctions $u_{n_1 n_2 m}$ that can be related to oscillator wave functions, so operators are defined by their effect on the eigenstates $|nfg\rangle$. Equations (5.8) together with $F_z|nfg\rangle = f|nfg\rangle$ and $G_z|nfg\rangle = g|nfg\rangle$ give the invariance subalgebra. An energy lowering operator Q which reduces n by 1 must reduce $F = \frac{1}{2}n - \frac{1}{2}$ by $\frac{1}{2}$ and so must change g and f by half-integers. If f and g are raised by $\frac{1}{2}$, Q must give zero when f or g is F, and so must multiply by a numerical factor containing $F - f$ and $F - g$ to some positive powers. The power $\frac{1}{2}$ is suggested by (5.8), and also by this giving the simplest eigenvalues $(F - f)(F - g)$ for $Q^\dagger Q$ which will be diagonal relative to the $|nfg\rangle$. Similar arguments indicate the other numerical factors in the following definitions (BK67) of Q_0, P_0, and $Q_\pm$:

$$
\begin{aligned}
Q_0|nfg\rangle &= (F+f)^{1/2}(F-g)^{1/2}|n-1 \ f-\tfrac{1}{2} \ g+\tfrac{1}{2}\rangle \\
P_0|nfg\rangle &= (F-f)^{1/2}(F+g)^{1/2}|n-1 \ f+\tfrac{1}{2} \ g-\tfrac{1}{2}\rangle \\
Q_-|nfg\rangle &= (F-f)^{1/2}(F-g)^{1/2}|n-1 \ f+\tfrac{1}{2} \ g+\tfrac{1}{2}\rangle \\
Q_+|nfg\rangle &= (F+f)^{1/2}(F+g)^{1/2}|n-1 \ f-\tfrac{1}{2} \ g-\tfrac{1}{2}\rangle
\end{aligned}
\tag{6.7}
$$

Their domain is the set of all bound states. As in Sections 1.10 and 6.3, these definitions imply the following equations for the complex conjugates:

$$
\begin{aligned}
Q_0^\dagger|nfg\rangle &= (F+f+1)^{1/2}(F-g+1)^{1/2}|n+1 \ f+\tfrac{1}{2} \ g-\tfrac{1}{2}\rangle \\
P_0^\dagger|nfg\rangle &= (F-f+1)^{1/2}(F+g+1)^{1/2}|n+1 \ f-\tfrac{1}{2} \ g+\tfrac{1}{2}\rangle \\
Q_-^\dagger|nfg\rangle &= (F-f+1)^{1/2}(F-g+1)^{1/2}|n+1 \ f-\tfrac{1}{2} \ g-\tfrac{1}{2}\rangle \\
Q_+^\dagger|nfg\rangle &= (F+f+1)^{1/2}(F+g+1)^{1/2}|n+1 \ f+\tfrac{1}{2} \ g+\tfrac{1}{2}\rangle
\end{aligned}
\tag{6.8}
$$

The action of the shift operators is illustrated in Figure 3. Also define N by $N|nfg\rangle = n|nfg\rangle$. Then evaluating commutators shows that the 15 operators form a Lie algebra. Forty-four of the 105 commutators are given in the accompanying table. The first entry means $[F_z, Q_0] = -\frac{1}{2}Q_0$, etc. By taking the

	Q_0	P_0	Q_-	Q_+
F_z	$-\tfrac{1}{2}Q_0$	$\tfrac{1}{2}P_0$	$\tfrac{1}{2}Q_-$	$-\tfrac{1}{2}Q_+$
F_+	$-Q_-$	0	0	$-P_0$
F_-	0	$-Q_+$	$-Q_0$	0
G_z	$\tfrac{1}{2}Q_0$	$-\tfrac{1}{2}P_0$	$\tfrac{1}{2}Q_-$	$-\tfrac{1}{2}Q_+$
G_+	0	$-Q_-$	0	$-Q_0$
G_-	$-Q_+$	0	$-P_0$	0
N	$-Q_0$	$-P_0$	$-Q_-$	$-Q_+$
Q	$G_z - F_z - N$	0	$-F_+$	$-G_-$
$P_0^\dagger$	0	$F_z - G_z - N$	$-G_+$	$-F_-$
$Q_-^\dagger$	$-F_-$	$-G_-$	$F_z + G_z - N$	0
$Q_+^\dagger$	$-G_+$	$-F_+$	0	$-F_z - G_z - N$

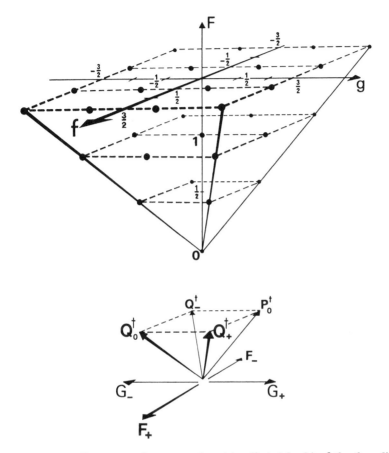

Figure 3.   Shift operators for energy eigenstates (first 4 levels) of the three-dimensional Coulomb problem.

complex conjugates of these equations, the first seven rows of the table give 28 more commutators, for example $[Q_0^\dagger, F_z] = -\frac{1}{2}Q_0^\dagger$. The 15 commutators of the invariance algebra were given in Section 3.3. The remaining 18 commutators are zero: N commutes with $\mathbf{F}$ and $\mathbf{G}$, the operators (6.7) mutually commute, and the operators (6.8) mutually commute.

This algebra has dimension 15, a $c*o_4$ subalgebra, and an element N which commutes with this subalgebra. These facts suggest that it is a $c*o_6$ algebra. The subset of skew-Hermitian operators should then be one of the real algebras o_6, $o_{5,1}$, or $o_{4,2}$, but not $o_{3,3}$ as the invariance subalgebra is o_4. To identify the algebra, first take combinations of the first six rows of the table above to get the commutators of (6.7) and the components of $\mathbf{F} \pm \mathbf{G}$. The results

	Q_0	P_0	Q_-	Q_+
L_z	0	0	Q_-	$-Q_+$
L_+	$-Q_-$	$-Q_-$	0	$-Q_0 - P_0$
L_-	$-Q_+$	$-Q_+$	$-Q_0 - P_0$	0

show that $Q_0 - P_0$ commutes with $\mathbf{L} = \mathbf{F} + \mathbf{G}$, and that $Q_\pm$ and $Q_0 + P_0$ are components of a vector operator with respect to $\mathbf{L}$. A standard basis for the algebra, as in Appendix D, therefore uses $Q_0 \pm P_0$ rather than Q_0 and P_0. Also, the Cartesian components of the vector operator involve $Q_+ \pm Q_-$. Finally, Hermitian (or skew-Hermitian) combinations such as $Q_0 + P_0 + Q_0^\dagger + P_0^\dagger$ must be used. Evaluating the commutators of these operators shows that Table 2 of Appendix D can be obtained with the following definitions: $T = N$ and

$$
\begin{aligned}
V_1 &= \tfrac{1}{2}(Q_+ + Q_+^\dagger - Q_- - Q_-^\dagger), & W_1 &= \tfrac{1}{2}i(Q_- - Q_-^\dagger - Q_+ + Q_+^\dagger) \\
V_2 &= \tfrac{1}{2}i(Q_+ - Q_+^\dagger + Q_- - Q_-^\dagger), & W_2 &= \tfrac{1}{2}(Q_- + Q_-^\dagger + Q_+ + Q_+^\dagger) \\
V_3 &= -\tfrac{1}{2}(Q_0 + Q_0^\dagger + P_0 + P_0^\dagger), & W_3 &= \tfrac{1}{2}i(Q_0 - Q_0^\dagger + P_0 - P_0^\dagger) \\
V_4 &= -\tfrac{1}{2}i(Q_0 - Q_0^\dagger - P_0 + P_0^\dagger), & W_4 &= -\tfrac{1}{2}(Q_0 + Q_0^\dagger - P_0 - P_0^\dagger)
\end{aligned}
$$

$$(6.9)$$

As in Section 4.3, $\mathbf{M} = \mathbf{F} - \mathbf{G}$. Thus the noninvariance algebra is an $o_{4,2}$ algebra. This result was first obtained in momentum space (MM65). The compact generator N has only integer eigenvalues, so that the noninvariance group is $SO_{4,2}$.

Matrix multiplication can be used to obtain the commutators of (6.9) from those given previously. Suppose X is an $n \times n$ skew-symmetric matrix ($\tilde{X} = -X$) with elements $[X_i, X_j]$, and $Y_\kappa = \sum_j p_{\kappa j} X_j$ where the $p_{\kappa j}$ are numbers forming an $m \times n$ matrix P. Then the commutators $[X_i, Y_\kappa]$ are the elements of the matrix $X\tilde{P}$, and the commutators $[Y_\kappa, Y_l]$ are the elements of the matrix $PX\tilde{P}$.

In Section 5.5 the shift operators $F_\pm$ and $G_\pm$ of the invariance algebra were related to products of oscillator shift operators, by using the relation

$$(\alpha\pi/n)^{1/2} R_{n'm}(\xi) R_{n''m}(\eta) e^{im\phi} = u_{n_1 n_2 m}(\xi, \eta, \phi) = (-)^{F-f} |nfg\rangle \quad (6.10)$$

between wave functions, where $n' = n - g + f$, $n'' = n + g - f$, and $\alpha = P/nh$. Equations like (5.26) can now be obtained for the operators of the noninvariance algebra. The effect of the oscillator operators has been given at the end of Section 6.2 ($M = \frac{1}{2}m$):

$$A_- R_{nm} e^{im\phi} = \mp(\tfrac{1}{2}n - \tfrac{1}{2}m - \tfrac{1}{2})^{1/2}\, R_{n-1\ m+1}\, e^{im\phi + i\phi}$$
$$A_+ R_{nm} e^{im\phi} = \pm(\tfrac{1}{2}n + \tfrac{1}{2}m - \tfrac{1}{2})^{1/2}\, R_{n-1\ m-1}\, e^{im\phi - i\phi}$$
$$A_-^* R_{nm} e^{im\phi} = \mp(\tfrac{1}{2}n - \tfrac{1}{2}m + \tfrac{1}{2})^{1/2}\, R_{n+1\ m-1}\, e^{im\phi - i\phi} \tag{6.11}$$
$$A_+^* R_{nm} e^{im\phi} = \pm(\tfrac{1}{2}n + \tfrac{1}{2}m + \tfrac{1}{2})^{1/2}\, R_{n+1\ m+1}\, e^{im\phi + i\phi}$$

with the negative sign for $m = 0$, the upper sign for $m > 0$, and the lower sign for $m < 0$. In the application to (6.10), R_{nm} will be $R_{n'm}(\xi)$. For the application of $B_\pm$ and $B_\pm^*$, R_{nm} will be $R_{n''m}(\eta)$.

The wave functions are eigenfunctions of $A_\pm^* A_\pm$ belonging to the eigenvalues $\frac{1}{2}(n - 1 - g + f \pm m)$, and are eigenfunctions of $B_\pm^* B_\pm$ belonging to the eigenvalues $\frac{1}{2}(n - 1 + g - f \pm m)$. As $A_+^* A_+ - A_-^* A_- = B_+^* B_+ - B_-^* B_-$ $(= m)$ on every wave function, only three of these operators are independent. Since $m = f + g$, $A_+^* A_+ = \frac{1}{2}N - \frac{1}{2} + F_z$, $A_-^* A_- = \frac{1}{2}N - \frac{1}{2} - G_z$, $B_+^* B_+ = \frac{1}{2}N - \frac{1}{2} + G_z$, and $B_-^* B_- = \frac{1}{2}N - \frac{1}{2} - F_z$.

Six possible energy lowering operators may be formed from products of pairs of the $A_\pm$ and $B_\pm$. $A_\pm^{\,2}$ change m by ± 2, and so do not appear in (6.7). Now take, for example, $A_+ e^{i\phi}B_+$. From (6.10) and (6.11),

$$S'(A_+ e^{i\phi}B_+)(-)^{F-f}\ n|nfg\rangle$$
$$= (F+f)^{1/2}(F+g)^{1/2}(-)^{F-f}\ (n-1)|n-1\ f-\tfrac{1}{2}\ g-\tfrac{1}{2}\rangle$$

in which $\alpha = P/nh$ and $n = 2F + 1$ have been used. As in the preceding section, a scaling operator S' is necessary because the oscillator frequency constant depends on the Coulomb energy. Comparison with (6.7) shows that $A_+ e^{i\phi}B_+$ corresponds to Q_+. A similar treatment of the other three cases gives the results

$$Q_0 = -SA_+ A_-, \qquad P_0 = SB_+ B_-, \qquad Q_\pm = \pm SA_\pm e^{\pm i\phi}B_\pm \tag{6.12}$$

in which $S = (n/n - 1)S'$. For Q_0 and P_0 the signs in (6.11) give -1, but for $Q_\pm$ they give $+1$ because $A_\pm$ and $B_\pm$ act on functions with the same m. The sign of $(-)^{F-f}$ must also be considered.

Using the explicit expressions (5.25) for the oscillator operators then leads to explicit expressions for the operators (6.7) and (6.8). Such expressions (for an $o_{4,1}$ subalgebra) have also been deduced (P66) by using the recurrence formulas for the Laguerre polynomials (i.e., confluent hypergeometric functions) which appear in the wave functions (5.3.) This approach has also been used (M66) to construct expressions for the generators of an $o_{4,1}$ subalgebra defined in terms of shift operators for the wave functions (5.1).

6.5. PROPERTIES OF SCALING OPERATORS

For a one-dimensional system with wave functions $\psi(x)$, the scaling operators $S(\beta)$ are defined by $S(\beta)\psi(x) = \beta^{1/2}\psi(\beta x)$, with $\beta > 0$. Obviously

$$S(\alpha)S(\beta) = S(\alpha\beta),$$

and if $V(x)$ is an operator, $S(\beta)V(x) = V(\beta x)S(\beta)$. The differentiation operator D gives $D\psi(\beta x) = \beta\psi'(\beta x)$, showing that $DS(\beta) = \beta S(\beta)D$. Taking $V(x) = x$, $xDS(\beta) = \beta xS(\beta)D = S(\beta)xD$. Thus $S(\beta)$ commutes with xD and should therefore be expressible as a function of xD.

The factor $\beta^{1/2}$ in the definition makes $S(\beta)$ unitary, because

$$\int \bar{\phi}(x)S(\beta)\psi(x)\,dx = \int \bar{\phi}(y/\beta)\beta^{1/2}\psi(y)\beta^{-1}\,dy$$

and $\beta^{-1/2}\phi(y/\beta) = S(1/\beta)\phi(y)$, showing that the complex conjugate of $S(\beta)$ is $S(1/\beta)$.

To find $S(\beta)$ as a function of xD, differentiate the definition with respect to β to get $S'(\beta)\psi(x) = \frac{1}{2}\beta^{-1/2}\psi(\beta x) + \beta^{1/2}x\psi'(\beta x)$, or

$$S'(\beta) = \tfrac{1}{2}\beta^{-1}S(\beta) + xS(\beta)D = \beta^{-1}(\tfrac{1}{2} + xD)S(\beta).$$

Since xD commutes with $S(\beta)$, this differential equation for $S(\beta)$ can be integrated to give

$$\log S(\beta) = (\tfrac{1}{2} + xD)\log\beta, \; S(\beta) = e^{(1/2 + xD)\log\beta} = \beta^{1/2}e^{xD\log\beta} \quad (6.13)$$

This demonstrates a simple example of a group obtained by an exponential function on an algebra. The operator $\frac{1}{2} + xD$ is skew-Hermitian; the operators $(\frac{1}{2} + xD)\log\beta$ are a real algebra of dimension 1; the $S(\beta)$ are a group of unitary operators. The group is noncompact, since the domain of the parameter $\log\beta$ is not finite.

For functions of n variables x_i, the definition becomes $S(\beta)\psi(\mathbf{x}) = \beta^{\frac{1}{2}n}\psi(\beta\mathbf{x})$, the numerical factor making $S(\beta)$ unitary if the functions are normalized with respect to the "volume element" $c\,dx_1\,dx_2\cdots dx_n$. Then $S(\beta)$ has similar properties, and indeed may be regarded as a product of n commuting operators which scale each variable in turn. For $n = 2$ the results are $(x_1 = u, x_2 = v)$:

$$S(\beta) = \exp\left[\left(1 + u\frac{\partial}{\partial u} + v\frac{\partial}{\partial v}\right)\log\beta\right] \quad (6.14)$$

$$S(\beta)V(u, v) = V(\beta u, \beta v)S(\beta), \; DS(\beta) = \beta S(\beta)D \quad (6.15)$$

in which D can be $\partial/\partial u$ or $\partial/\partial v$.

The virial theorem may be deduced (L59) from these results on scaling operators. As the wave function ϕ is varied, the energy expectation values

$\int \bar{\phi} H \phi \, d\tau$ are stationary when ϕ is an eigenfunction. So if ψ is an energy eigenfunction, putting $\phi = S(\beta)\psi$ gives $d/d\beta(\int \bar{\psi} S(1/\beta) HS(\beta)\psi \, d\tau) = 0$ at $\beta = 1$, since $S(1/\beta)$ is the complex conjugate of $S(\beta)$. From (6.15), $D^2 S(\beta) = \beta^2 S(\beta) D^2$, in which D^2 can be replaced by the kinetic energy T. Also from (6.15),

$$S(\beta) V(\mathbf{x}/\beta) = V(\mathbf{x}) S(\beta).$$

Thus $S(1/\beta) HS(\beta) = \beta^2 T + V(\mathbf{x}/\beta)$.

Now $V(\mathbf{x}/\beta) = V(y_1, \ldots, y_n)$ with $y_i = x_i/\beta$, and $d/d\beta \, V(\mathbf{x}/\beta)$ is

$$\sum_i V_i \frac{dy_i}{d\beta} = -\frac{1}{\beta^2} \sum_i x_i V_i(\mathbf{x}/\beta)$$

where the V_i are the partial derivatives of V. Putting $\beta = 1$ gives the virial theorem: $2 \int \bar{\psi} T\psi \, d\tau = \int \bar{\psi}(\sum x_i V_i)\psi \, d\tau$. When V is a homogeneous function of the x_i of degree k, Euler's theorem says that $\sum x_i V_i = kV$. For the harmonic oscillator, $k = 2$, and $\langle T \rangle = \langle V \rangle$ in energy eigenstates. For the Coulomb problem, $k = -1$, and $2\langle T \rangle + \langle V \rangle = 0$, a result previously obtained in Section 3.6.

6.6. SCALING OPERATORS AND THE TWO-DIMENSIONAL OSCILLATOR

Substituting explicit expressions for the shift operators into V_3 given in (6.3), and comparing with (6.14), shows that $S(\beta) = \exp(2iV_3 \log \beta)$. Since $[V_3, J_3] = 0$, $[S(\beta), J_3] = 0$, and the matrix elements of $S(\beta)$ between states of different angular momentum are zero. This is also obvious because the definition of the scaling operator means that angular coordinates are unchanged. In this section the matrix element $\langle Nm | S(\beta) | nm \rangle$ will be obtained using the $o_{3,2}$ noninvariance algebra.

From Table 1 of Appendix D, U_3, V_3, and S span an $o_{2,1}$ subalgebra $\mathcal{M}$. Every operator of $\mathcal{M}$ commutes with L_3, and so any set of states with the same angular momentum is invariant under $\mathcal{M}$. All the states of the system form the domain of a reducible representation of $o_{2,1}$, the representatives being the operators of $\mathcal{M}$. Choosing eigenstates $|nm\rangle$ of L_3 as a basis shows the decomposition into irreducible representations which may be labeled by the eigenvalues m of L_3. Any scaling operator $S(\beta)$ is an element of the group generated by $\mathcal{M}$, so it should be possible to get the matrix elements from representation theory.

The correspondence $S \leftrightarrow T_z$, $U_3 \leftrightarrow T_x$, $V_3 \leftrightarrow T_y$ exhibits the standard Hermitian basis of Section 2.5. The compact generator S is the Hamiltonian divided by $2\hbar\omega$, hence the eigenvalues of S are $\frac{1}{2}n$. These half-integral eigenvalues imply the corresponding group is $SU_{1,1}$: the scaling operators obtained by taking

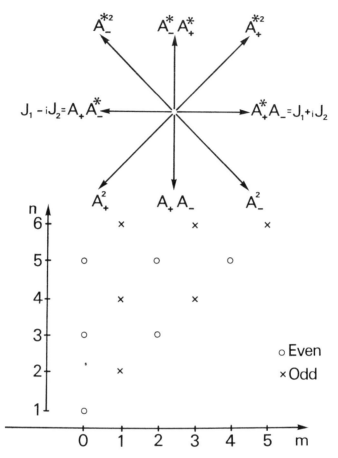

Figure 4. Shift operators for two-dimensional oscillator eigenstates of energy (first 6 levels) and angular momentum (positive).

all real β are a subgroup of an $SU_{1,1}$ group. In the representation obtained from the states $|nm\rangle$ with a fixed value of m, the eigenvalues of S are $\frac{1}{2}|m| + \frac{1}{2}$, $\frac{1}{2}|m| + \frac{3}{2}$, ... From Section 2.5 the representation is type D_k^+ with $k = \frac{1}{2}|m| + \frac{1}{2}$ and $a = \frac{1}{2}(n - |m| - 1)$; the value of the Casimir operator is $k - k^2 = \frac{1}{4}(1 - m^2)$. States of arbitrarily high energy can be obtained using the raising operator $U_3 + iV_3$ because a representation of type D_k^+ is irreducible and not finite dimensional. Two representations with m having opposite sign are isomorphic (k and a depend only on $|m|$), so that it is sufficient to consider $m \geq 0$. The required matrix elements therefore depend on m

through $|m|$ only, which is consistent with the same property of $R_{nm}(\xi)$, the part of the wave function changed by a scaling operator. For any β, $S(\beta)|nm\rangle$ also belongs to the domain $\mathscr{S}_k$ of the representation D_k^+, and is orthogonal to $|NM\rangle$ of $\mathscr{S}_K$ if $k \neq K$, that is, $m \neq M$. Figure 4 illustrates some of the states with m positive, and their quantum numbers, together with the action of the shift operators. The states on each vertical line span the domain of an irreducible representation.

The group generated by V_3 corresponds to the group generated by T_y in Section 2.5, so (2.19) and (2.20) give the required matrix elements on putting $\lambda = 2 \log \beta$. Then $\sinh \frac{1}{2}\lambda = \frac{1}{2}\beta - \frac{1}{2}\beta^{-1}$, $\cosh \frac{1}{2}\lambda = \frac{1}{2}\beta + \frac{1}{2}\beta^{-1}$, and $\coth \frac{1}{2}\lambda = (\beta^2 + 1)/(\beta^2 - 1)$, so that $\langle Nm|S(\beta)|nm\rangle$ is $(2k = m + 1)$

$$\binom{a + m}{a}^{1/2} \binom{A + m}{A}^{1/2} \frac{(-)^A (2\beta)^{m+1} (\beta^2 - 1)^{a+A}}{(\beta^2 + 1)^{a+A+m+1}}$$

$$\times {}_2F_1\left[-a, -A; m + 1; \frac{-4\beta^2}{(\beta^2 - 1)^2}\right] \quad (6.16)$$

where $a = \frac{1}{2}n - \frac{1}{2} - \frac{1}{2}m$ and $A = \frac{1}{2}N - \frac{1}{2} - \frac{1}{2}m$. Here $m \geq 0$, but can be replaced by $|m|$.

This independence of the sign of m can also be seen by the method used in Section 2.3. If $R = \exp(-i\pi J_1)$, then $RV_3 = -V_3 R$, $RV_3^s = (-V_3)^s R$, and so $S(\beta)R = \exp(2iV_3 \log \beta) R = R \exp(-2iV_3 \log \beta) = RS(1/\beta)$; also $R|nm\rangle = i^{n-1} |n - m\rangle$. Then

$$\langle N - m|S(\beta)|n - m\rangle = i^{-n+N}\langle Nm|S(1/\beta)|nm\rangle$$
$$= (-)^{N-m-1}\langle Nm|S(\beta)|nm\rangle,$$

using (6.16), and $N - m - 1$ is even.

Now (2.19) is only correct for states satisfying (2.18) with $p = -1$. Since $T_- \leftrightarrow U_3 - iV_3 = -A_+A_-$, (6.16) is only correct if

$$A_+ A_-|nm\rangle = \frac{1}{2}(n - m - 1)^{1/2}(n + m - 1)^{1/2}|n-2 \ m\rangle \quad (6.17)$$

From the remarks at the end of Section 6.2, the wave functions ψ_{nM} $(M = \frac{1}{2}m)$ are a realization of such states, and so (6.16) is $\int_0^\infty \int_0^{2\pi} \frac{1}{2} \, d\xi \, d\phi \, \bar{\psi}_{NM} \, S(\beta)\psi_{nM}$. However, (6.11) shows that the wave functions $R_{nm}e^{im\phi}$ give (2.18) with $p = 1$. From the remark after (2.20) [or as the ψ_{nM} include the phase factor $(-)^a$], changing $(\beta^2 - 1)$ to $(1 - \beta^2)$ in (6.16) gives $\int_0^\infty \int_0^{2\pi} \frac{1}{2} \, d\xi \, d\phi \bar{R}_{Nm} S(\beta)R_{nm}$ as calculated by integration (BW66). Since $\xi = u^2 + v^2$, (6.14) becomes

$$\exp\left[\left(1 + 2\xi\frac{\partial}{\partial \xi}\right) \log \beta\right] = \beta \exp\left[\xi \frac{\partial}{\partial \xi} \log \beta^2\right],$$

and therefore $S(\beta)R(\xi) = \beta R(\beta^2 \xi)$.

6.7. THE OSCILLATOR REPRESENTATION OF $o_{4,2}$

Explicit expressions for the operators of the Coulomb noninvariance algebra contain scaling operators and renormalization factors. The noninvariance algebra of the two-dimensional problem is isomorphic to an algebra defined on the even parity states of the two-dimensional oscillator. Explicit forms of the operators of this isomorphic algebra are simpler, since they are just products of oscillator shift operators. This makes it convenient to consider algebraic aspects of the Coulomb problem in terms of the isomorphic oscillator algebra. For the three-dimensional problem, the operators of the isomorphic algebra are obtained from those of Section 6.4 by putting $S = 1$. Although these operators were found by considering the two-dimensional oscillator, their domain does not consist of the wave functions of a physical oscillator. The normalization in the domain may therefore be defined in any convenient manner.

Consider the domain spanned by the functions $R_{Sm}(\xi)R_{Tm}(\eta)e^{im\phi}$, where S and T are integers greater than $|m|$ and having the opposite parity to the integer m. Define an inner product by integration with respect to the " volume element " $d\xi\,d\eta\,d\phi$. The normalized function

$$(-)^{\frac{1}{2}(T-m-1)}(\tfrac{1}{2}\pi)^{1/2}R_{Sm}(\xi)R_{Tm}(\eta)\ e^{im\phi} \tag{6.18}$$

will be denoted by $|STm\rangle$. Its phase has been chosen to correspond to that of $|nfg\rangle$, using (5.21) and (5.24).

On this domain the operators $A_{\pm}^{*}\,A_{\pm}$, $B_{\pm}^{*}\,B_{\pm}$, $A_{+}^{*}\,e^{-i\phi}\,B_{-}$, $A_{-}\,e^{-i\phi}\,B_{+}^{*}$, $A_{+}\,A_{-}$, $B_{+}\,B_{-}$, $A_{\pm}\,e^{\pm i\phi}\,B_{\pm}$, $A_{+}e^{i\phi}\,B_{-}^{*}$, $A_{-}^{*}\,e^{i\phi}\,B_{+}$, $A_{+}^{*}\,A_{-}^{*}$, $B_{+}^{*}B_{-}^{*}$, $A_{\pm}^{*}\,e^{\mp i\phi}\,B_{\pm}^{*}$ are defined by the explicit expressions (5.25) for the oscillator shift operators. The first four are Hermitian (but not independent), and the next six are complex conjugates of the last six. From (6.11) the following matrix elements (BK67, BK67a, K68, F69) can be written down:

$$\langle S-2\ T\ m|A_{+}\,A_{-}|S\ T\ m\rangle\quad = -\tfrac{1}{2}(S-m-1)^{1/2}(S+m-1)^{1/2}$$

$$\langle S\ T-2\ m|B_{+}\,B_{-}|S\ T\ m\rangle\quad = \tfrac{1}{2}(T-m-1)^{1/2}(T+m-1)^{1/2}$$

$$\langle S+2\ T\ m|A_{+}^{*}\,A_{-}^{*}|S\ T\ m\rangle\quad = -\tfrac{1}{2}(S+m+1)^{1/2}(S-m+1)^{1/2}$$

$$\langle S\ T+2\ m|B_{+}^{*}\,B_{-}^{*}|S\ T\ m\rangle\quad = \ \ \tfrac{1}{2}(T+m+1)^{1/2}(T-m+1)^{1/2}$$

$$\langle S+1\ T+1\ m\pm1|A_{\pm}^{*}\,e^{\mp i\phi}\,B_{\pm}^{*}|S\ T\ m\rangle = \pm\tfrac{1}{2}(S\pm m+1)^{1/2}(T\pm m+1)^{1/2}$$

$$\langle S-1\ T-1\ m\mp1|A_{\pm}\,e^{\pm i\phi}\,B_{\pm}|S\ T\ m\rangle = \pm\tfrac{1}{2}(S\pm m-1)^{1/2}(T\pm m-1)^{1/2}$$

$$\tag{6.19}$$

$$\langle S\ T\ m\ |A_{\pm}^{*}\,A_{\pm}|S\ T\ m\rangle\quad = \ \ \tfrac{1}{2}(S\pm m-1)$$

$$\langle S\ T\ m\ |B_{\pm}^{*}\,B_{\pm}|S\ T\ m\rangle\quad = \ \ \tfrac{1}{2}(T\pm m-1)$$

$$\langle S+1\ T-1\ m\mp1|A_{\mp}^{*}\,e^{\pm i\phi}\,B_{\pm}|S\ T\ m\rangle = \mp\tfrac{1}{2}(S\mp m+1)^{1/2}(T\pm m-1)^{1/2}$$

$$\langle S-1\ T+1\ m\mp1|A_{\pm}\,e^{\pm i\phi}B_{\mp}^{*}|S\ T\ m\rangle = \pm\tfrac{1}{2}(S\pm m-1)^{1/2}(T\mp m+1)^{1/2}$$

The first six of these equations, which have already been used to establish (6.12), correspond to (6.7) and (6.8), and so the work in Section 6.4 may be reinterpreted to give the commutators of these operators. Hence the Hermitian linear combinations listed in Appendix E have the commutators given in Table 2 of Appendix D. This will be called the oscillator representation of $o_{4,2}$.

In this representation, the functions $|STm\rangle$ all have the same oscillator constant α, which appears in R_{Sm} and R_{Tm} as in Section 5.5. To change $|STm\rangle$ to the corresponding hydrogen atom eigenfunction $|nfg\rangle$ requires replacing α by $P/nh = 2P/h(S+T)$, and this can be done with a scaling operator. Define $D(\beta)$ to change ξ and η to $\beta\xi$ and $\beta\eta$, and multiply by β. From the end of the preceding section, $D(\beta)$ is just a product of two operators $S(\beta^{1/2})$, one changing ξ and the other changing η; explicitly, and using (E.2) :

$$D(\beta) = \beta \exp[(\xi\, \partial/\partial\xi + \eta\, \partial/\partial\eta) \log \beta] \leftrightarrow \exp(-iV_4 \log \beta).$$

The matrix element $\langle \sigma\tau m| D(\beta)| STm\rangle$ is therefore a product of two expressions like (6.16). Putting

$$\chi = \tfrac{1}{2}\sigma - \tfrac{1}{2} - \tfrac{1}{2}|m|, \; \zeta = \tfrac{1}{2}\tau - \tfrac{1}{2} - \tfrac{1}{2}|m|, \; X = \tfrac{1}{2}S - \tfrac{1}{2} - \tfrac{1}{2}|m|,$$

and $Z = \tfrac{1}{2}T - \tfrac{1}{2} - \tfrac{1}{2}|m|$, gives

$$\langle \sigma\tau m| D(\beta)| STm\rangle$$

$$= \left[\frac{(\chi+|m|)!\,(\zeta+|m|)!\,(X+|m|)!\,(Z+|m|)!}{\chi!\,\zeta!\,X!\,Z!(|m|!)^4}\right]^{1/2} \left[\frac{4\beta}{(1-\beta)^2}\right]^{|m|+1}$$

$$\times \left[\frac{1-\beta}{1+\beta}\right]^{\frac{1}{2}(\sigma+\tau+S+T)}$$

$$\times (-)^{\frac{1}{2}\sigma+\frac{1}{2}T+\tau} F\left[-\chi, -X; |m|+1; \frac{-4\beta}{(1-\beta)^2}\right] \qquad (6.20)$$

$$\times F\left[-\zeta, -Z; |m|+1; \frac{-4\beta}{(1-\beta)^2}\right]$$

The phase factor in (6.20) is a combination of four terms—two from the factor in (6.18), and $(-)^\chi(-)^\zeta$ from the $(-)^A$ appearing in (6.16). The (β^2-1) in (6.16) has been changed to $(1-\beta^2)$ according to the remark on phase at the end of Section 6.6. Since σ, τ, S, and T all have the opposite parity to m, $(-)^\sigma = (-)^\tau = (-)^S = (-)^{m+1} = (-)^T$. The result $D^*(\beta) = D(1/\beta)$ is then obvious from (6.20).

In Section 6.9 $\langle \sigma\tau m| D(\beta)| STm\rangle$ is obtained without reference to $S(\beta)$.

6.8. MATRIX ELEMENTS OF POSITION AND MOMENTUM

From (6.10) and (6.18) the hydrogen atom eigenstates are

$$|nfg\rangle = \frac{1}{n}\left[\frac{2P}{h}\right]^{1/2} D\left(\frac{P}{nh\alpha}\right)|STm\rangle$$

with $S = n - g + f$ and $T = n + g - f$. The $|nfg\rangle$ are normalized with respect to the volume element $\frac{1}{4}(\xi + \eta)\, d\xi\, d\eta\, d\phi$. Matrix elements for the hydrogen atom are therefore related to those of the oscillator representation by

$$\langle n'f'g'|A|nfg\rangle = \frac{P}{2nhn'}\langle \sigma\tau m'|D^*\left(\frac{P}{n'h\alpha}\right)(\xi + \eta)A D\left(\frac{P}{nh\alpha}\right)|STm\rangle$$

(6.21)

where $m' = f' + g'$, $\sigma = n' - g' + f'$, and $\tau = n' + g' - f'$.

Taking $A = -hP/\mu r = -2hP/\mu(\xi + \eta)$ gives

$$-(P^2/\mu nn')\langle \sigma\tau m'|D(n'/n)|STm\rangle$$

for the potential energy matrix elements, which can then b written down from (6.20). The case $n = n'$ gives $\delta_{\sigma,S}\delta_{\tau,T}\delta_{m',m}(-P^2/\mu n^2)$ or $\delta_{f',f}\delta_{g',g}2E_n$, which is the virial theorem again.

Next take $A = p_x = -ih\partial/\partial x$. The results in Appendix E show that in the oscillator representation $-ih(\xi + \eta)\partial/\partial x \leftrightarrow 2hW_1$, and so commutes with $D(\beta) \leftrightarrow \exp(-iV_4 \log \beta)$. So

$$\langle n'f'g'|\mathbf{p}|nfg\rangle = \frac{P}{2inn'}\sum_{S'T'}\langle \sigma\tau m'|D\left(\frac{n'}{n}\right)|S'T'm'\rangle\langle S'T'm'|(\xi + \eta)\mathbf{V}|STm\rangle$$

Again the matrix elements of $D(n'/n)$ are given in (6.20); and the matrix elements of $(\xi + \eta)\partial/\partial x \leftrightarrow 2iW_1$ can be written down from (E.1) and (6.19). For p_x and the case $m' = m - 1$, (6.19) shows that there are only two terms in the sum over $S'T'$: $S' = S + 1, T' = T + 1$ from $-A_-^* e^{i\phi}B_-^*$ and $S' = S - 1, T' = T - 1$ from $A_+ e^{i\phi}B_+$. Using (6.19) gives ($f' + g' = m - 1$):

$$\frac{4inn'}{P}\langle n'f'g'|p_x|nfg\rangle$$

$$= \langle \sigma\ \tau\ m-1|D\left(\frac{n'}{n}\right)|S+1\ T+1\ m-1\rangle\ (S - m + 1)^{1/2}(T - m + 1)^{1/2}$$

$$+\langle \sigma\ \tau\ m-1|D\left(\frac{n'}{n}\right)|S-1\ T-1\ m-1\rangle\ (S + m - 1)^{1/2}(T + m - 1)^{1/2}$$

Assuming $m \geq 1$, (6.20) must be used with $|m|$, S, and T replaced by $m - 1$, $S \pm 1$, and $T \pm 1$ respectively. The result is

$\langle n'f'g' | p_x | nfg \rangle$

$$= \frac{P(-)^{F-F'-f+g'}}{2inn'(m-1)!^2} \left[\frac{(F'+f')!(F'+g')!(F+f)!(F+g)!}{(F'-f')!(F'-g')!(F-f)!(F-g)!} \right]^{1/2}$$

$$\times \left[\frac{4nn'}{(n-n')^2} \right]^m \left[\frac{n-n'}{n+n'} \right]^{n+n'-1}$$

$$\times \left\{ {}_2F_1 \left[g' - F', g - F; m; \frac{-4nn'}{(n-n')^2} \right] \right. \tag{6.22}$$

$$\times {}_2F_1 \left[f' - F', f - F; m; \frac{-4nn'}{(n-n')^2} \right]$$

$$- \left[\frac{n-n'}{n+n'} \right]^2 {}_2F_1 \left[g' - F', g - F - 1; m; \frac{-4nn'}{(n-n')^2} \right]$$

$$\left. \times {}_2F_1 \left[f' - F', f - F - 1; m; \frac{-4nn'}{(n-n')^2} \right] \right\}$$

where $m = f + g = f' + g' + 1 \geq 1$, $F = \frac{1}{2}n - \frac{1}{2}$, and $F' = \frac{1}{2}n' - \frac{1}{2}$. The phase difference from $\langle n'g'f' | p_x | ngf \rangle$, derived in Section 3.6 by considering parity, follows from (6.22) on using $f' - g = f - g' - 1$ and $(-)^{2f} = (-)^{2F}$.

Similarly, using $(\xi + \eta)\partial/\partial z \leftrightarrow 2iW_3$, (E.1) and (6.19) show that $(m' = m)$

$\langle n'f'g' | p_z | nfg \rangle$ $(4inn'/P)$

$$= \langle \sigma\tau m | D\left(\frac{n'}{n}\right) | S+2 \ T \ m \rangle \ (S-m+1)^{1/2} \ (S+m+1)^{1/2}$$

$$+ \langle \sigma\tau m | D\left(\frac{n'}{n}\right) | S \ T+2 \ m \rangle \ (T-m+1)^{1/2} \ (T+m+1)^{1/2}$$

$$- \langle \sigma\tau m | D\left(\frac{n'}{n}\right) | S-2 \ T \ m \rangle \ (S-m-1)^{1/2} \ (S+m-1)^{1/2}$$

$$- \langle \sigma\tau m | D\left(\frac{n'}{n}\right) | S \ T-2 \ m \rangle \ (T-m-1)^{1/2} \ (T+m-1)^{1/2}$$

In the factors arising from (6.19), m can be replaced by $|m|$, and so the algebraic manipulations do not depend on the sign of m. A concise form of the result is obtained by using the quantum numbers n_1 and n_2 appearing in (5.3) and (5.16), and defining (G29)

$$\Psi_m(\lambda, v) = {}_2F_1[-\lambda, -v; |m| + 1; -4nn'/(n-n')^2].$$

Then

$$S - |m| - 1 = 2n_1, \ T - |m| - 1 = 2n_2, \ \sigma - |m| - 1 = 2n_1', \ \tau - |m| - 1 = 2n_2'$$

and

$$\langle n'f'g' | p_z | nfg \rangle$$

$$= \frac{P(-)^{n_1'+n_2}}{2inn'|m|!^2} \left[\frac{(n_1' + |m|)!(n_2' + |m|)!(n_1 + |m|)!(n_2 + |m|)!}{n_1'!n_2'!n_1!n_2!} \right]^{1/2}$$

$$\times \left[\frac{4nn'}{(n-n')^2} \right]^{|m|+1} \left[\frac{n-n'}{n+n'} \right]^{n+n'-1}$$

$$\times \left\{ n_2 \Psi_m(n_1', n_1) \Psi_m(n_2', n_2 - 1) - n_1 \Psi_m(n_1', n_1 - 1) \Psi_m(n_2', n_2) \right.$$

$$- \left[\frac{n-n'}{n+n'} \right]^2 [(n - n_1) \Psi_m(n_1', n_1) \Psi_m(n_2', n_2 + 1)$$

$$\left. - (n - n_2) \Psi_m(n_1', n_1 + 1) \Psi_m(n_2', n_2)] \right\} \qquad (6.23)$$

Since $n = n_1 + n_2 + |m| + 1$, the transformation formula (O65)

$$(c + v)F(-v', -v - 1; c; z) = v(z - 1)F(-v', -v + 1; c; z)$$
$$+ (c + 2v + v'z - vz)F(-v', -v; c; z)$$

between contiguous hypergeometric functions may be used (with $c = |m| + 1$ and $v = n_1$ and n_2) to write the { } brace in (6.23) as

$$2[n_2 \Psi_m(n_1', n_1) \Psi_m(n_2', n_2 - 1) - n_1 \Psi_m(n_1', n_1 - 1) \Psi_m(n_2', n_2)]$$

$$- \frac{2}{(n+n')^2} [2nn'(n_1' - n_2') + (n^2 + n'^2)(n_2 - n_1)] \Psi_m(n_1', n_1) \Psi_m(n_2', n_2)$$

$$(6.24)$$

According to (5.16) and (5.21), $\int \bar{u}_{n_1'n_2'm} p_z u_{n_1 n_2 m} d\mathbf{r}$ is obtained from (6.23) on multiplying by $(-)^{n_2+n_2'}$.

Matrix elements of $\mathbf{r}$ may now be obtained from those of $\mathbf{p}$ by multiplying by $ih/\mu(E_n - E_{n'}) = 2ihn^2n'^2/P^2(n^2 - n'^2)$. This hypervirial theorem was given in Section 3.6. Using (6.24) in (6.23), and multiplying by $(-)^{n_2+n_2'}$, gives $\int \bar{u}_{n_1'n_2'm} z u_{n_1 n_2 m} d\mathbf{r}$. The result contains the phase factor $(-)^{n_1'+n_2'}$ and agrees with (65.1) of Bethe and Salpeter (B57) on interchanging $n_1'n_2' n'$ with $n_1 n_2 n$ in their equation. Similarly $\int \bar{u}_{n_1'n_2'm-1} x u_{n_1 n_2 m} d\mathbf{r}$ follows from (6.22) in agreement with (65.2) of (B57) on putting $F - f = n_2$, $F - g = n_1$, $f + g = m$, $F' - f' = n_2'$, $F' - g' = n_1'$, and $f' + g' = m - 1$. Replacing m by $m + 1$ gives $\langle m | x | m + 1 \rangle$ which is $\langle m + 1 | x | m \rangle$ because x is Hermitian and the matrix element is real; as $\langle m - 1 | x + iy | m \rangle = 0$, $\langle m - 1 | y | m \rangle = -i \langle m - 1 | x | m \rangle$, and similarly $\langle m + 1 | y | m \rangle = i \langle m + 1 | x | m \rangle$.

Matrix elements of $\mathbf{r}$ can also be obtained without finding those of $\mathbf{p}$. From (E.2) and (E.3), the operators representing $(V_i + M_i)/\alpha$ $(i = 1, 2, 3)$ correspond to $x, y,$ and z in the Coulomb problem. It is convenient to let the symbols $L_1, L_2, \ldots, W_4, N$ denote the representatives in the oscillator representation, so that $(V_3 + M_3)/\alpha = \frac{1}{2}\xi - \frac{1}{2}\eta$, etc. Then also $(N + W_4)/\alpha = \frac{1}{2}\xi + \frac{1}{2}\eta$. If $A = \mathbf{r}$ in (6.21), it can be evaluated by putting in the right side $(\xi + \eta)A = 2(N + W_4)$ $(\mathbf{V} + \mathbf{M})/\alpha^2$. To simplify requires the commutation of the dilation operator $\exp(-iV_4 \log \beta)$ with $N + W_4$ and $\mathbf{V} + \mathbf{M}$. These operators just multiply, so that (6.15) gives

$$D(\beta)(N + W_4)(V_i + M_i) = \beta^2 (N + W_4)(V_i + M_i)D(\beta) \qquad (6.25)$$

and (6.21) becomes

$$\langle n'f'g' | \mathbf{r} | nfg \rangle = \left(\frac{n'h}{nP}\right) \left\langle \sigma\tau m' \left| (N + W_4)(\mathbf{V} + \mathbf{M}) \; D\left(\frac{n'}{n}\right) \right| STm \right\rangle$$
$$(6.26)$$

which can be evaluated as above using (6.19) and (6.20). For the case $n' = n$, only (6.19) is necessary, and equations (3.15) are thereby checked.

The commutation (6.25) can also be derived directly from the $o_{4,2}$ commutation relations. If $B = N + W_4$ or $V_i + M_i$, then $[V_4, B] = iB$, and so putting $A = V_4$ in (G64)

$$e^{aA}Be^{-aA} = B + a[A, B] + \tfrac{1}{2}a^2[A, [A, B]] + \cdots$$

gives $e^{ia} B$ and so $(a = -i \log \beta)$

$$D(\beta)B = \beta B D(\beta). \qquad (6.27)$$

Then $D(\beta)B^r = \beta^r B^r D(\beta)$, and so $D(\beta) \; e^{bB} = e^{\beta b B} \; D(\beta)$. $\qquad (6.28)$

6.9. TRANSITION FORM FACTORS

Substituting $(\xi + \eta)e^{ikz} = (2/\alpha)(N + W_4) \exp[ik(V_3 + M_3)/\alpha]$ in (6.21), using (6.27) with $B = N + W_4$ and using (6.28) with $B = V_3 + M_3$, gives

$$n\langle n'f'g' | e^{ikz} | nfg \rangle = \langle \sigma\tau m | (N + W_4)E | STm \rangle$$

where

$$E = \exp[in'hk(V_3 + M_3)/P] \exp[iV_4 \log(n/n')].$$

The arguments of these exponential functions are elements of the $o_{4,2}$ algebra, so that E is an element of the group generated (BK67a), and the matrix elements of E can be found algebraically. Since it is sufficient to consider an $o_{2,2}$

subalgebra (BK67b, K68), which is a direct sum of commuting $o_{2,1}$ algebras, the matrix elements can be found as in Section 1.13.

The standard Hermitian bases of the commuting $o_{2,1}$ algebras are $A_1 = -\frac{1}{2}W_4 - \frac{1}{2}V_3$, $A_2 = -\frac{1}{2}V_4 + \frac{1}{2}W_3$, $A_3 = \frac{1}{2}N + \frac{1}{2}M_3$, and $B_1 = \frac{1}{2}V_3 - \frac{1}{2}W_4$, $B_2 = -\frac{1}{2}V_4 - \frac{1}{2}W_3$, $B_3 = \frac{1}{2}N - \frac{1}{2}M_3$. From (E.1), the A_i are functions of $A_\pm$ and $A_\pm^*$ and the B_i are the same functions of $B_\pm$ and $B_\pm^*$. The A_i are just the operators U_3, V_3, and S of Sections 6.2 and 6.6 with their domain extended, and so give representations of $o_{2,1}$ of the type D_k^+ with $k = \frac{1}{2}|m| + \frac{1}{2}$ and $a = \frac{1}{2}S - \frac{1}{2}|m| - \frac{1}{2}$; similarly the B_i give representations with $a = \frac{1}{2}T - \frac{1}{2}|m| - \frac{1}{2}$. The space spanned by the functions $|STm\rangle$ with m and T(or S) fixed is invariant under the A_i (or the B_i).

Then $V_3 + M_3 = A_3 - A_1 - B_3 + B_1$ and $V_4 = -A_2 - B_2$, so $E = E_A E_B$ with

$$E_A = \exp[in'K(A_3 - A_1)] \exp(i\lambda A_2)$$

and $E_B = \exp[in'K(B_1 - B_3)] \exp(i\lambda B_2)$, where $K = hk/P$ and $\lambda = \log(n'/n)$. Since E_A does not change T and E_B does not change S, $\langle \sigma\tau m | E | STm \rangle = \langle \sigma\tau m | E_A | S\tau m \rangle \langle S\tau m | E_B | STm \rangle$.

To get $\langle \sigma\tau m | E_A | S\tau m \rangle$, the method for $\langle kA | \exp(i\lambda T_y) | ka \rangle$ in Sections 2.5 and 1.13 will be followed. Correspondences with the fundamental representation are

$$\exp(i\lambda A_2) \leftrightarrow I \cosh \frac{1}{2}\lambda - \sigma_2 \sinh \frac{1}{2}\lambda = \frac{(n+n')I + (n-n')\sigma_2}{2\sqrt{(nn')}}$$

$$A_3 - A_1 \leftrightarrow \frac{1}{2}\sigma_3 - \frac{1}{2}i\sigma_1, \ (A_3 - A_1)^2 \leftrightarrow 0, \ \exp[in'K(A_3 - A_1)]$$
$$\leftrightarrow I + \frac{1}{2}n'K(\sigma_1 + i\sigma_3)$$

$$E_A \leftrightarrow \frac{(n+n')I + (n-n')\sigma_2 + nn'K(\sigma_1 + i\sigma_3)}{2\sqrt{(nn')}}$$

Therefore the $SU_{1,1}$ transformation corresponding to E_A is

$$z \to \frac{z(n+n'+inn'K) + w(in - in' + nn'K)}{2\sqrt{(nn')}}$$

$$w \to \frac{z(in' - in + nn'K) + w(n+n' - inn'K)}{2\sqrt{(nn')}}$$

The next steps are as in Section 1.13, including the same manipulation of $\binom{a}{r}\binom{-a-2k}{A-r}$ to write the sum over r in terms of $_2F_1(-a, -A; -a - A - 2k + 1; 1 - y)$ with $y = -4nn'/\{(n-n')^2 + K^2 n^2 n'^2\}$. As in (2.20), a transformation formula leads to $_2F_1(-a, -A; 2k; y)$. As in Section 2.5, phase and normalization must be considered in setting up the correspondence with the representation in Section 1.12. Since $A_1 - iA_2 = -A_+ A_-$, $p = 1$ in (2.18), requiring

$$|STm\rangle \leftrightarrow \left(\frac{-b-1}{a}\right)^{1/2} z^a w^b i^b \qquad (a = \tfrac{1}{2}S - \tfrac{1}{2}|m| - \tfrac{1}{2}, b = -a - |m| - 1)$$

The result will be written with the notation used in (6.20): $\chi = \tfrac{1}{2}\sigma - \tfrac{1}{2} - \tfrac{1}{2}|m|$, $X = \tfrac{1}{2}S - \tfrac{1}{2} - \tfrac{1}{2}|m|$; and $|m|$ will be replaced by m. Then

$$\langle \sigma\tau m | E_A | S\tau m\rangle$$

$$= \left[\frac{(\chi+m)!(X+m)!}{\chi!\,m!\,X!\,m!}\right]^{1/2} \frac{(n-n'-iKnn')^X (n'-n-iKnn')^\chi}{(n+n'-iKnn')^{\frac{1}{2}S+\frac{1}{2}\sigma}}$$

$$\times (4nn')^{\frac{1}{2}m+\frac{1}{2}} \;{}_2F_1(-\chi, -X; m+1; y)$$

For $\langle S\tau m | E_B | S\tau m\rangle$ K is replaced by $-K$, and there is a phase difference because $B_1 - iB_2 = -B_+ B_-$ gives (2.18) with $p = -1$. With $\zeta = \tfrac{1}{2}\tau - \tfrac{1}{2} - \tfrac{1}{2}m$ and $Z = \tfrac{1}{2}T - \tfrac{1}{2} - \tfrac{1}{2}m$, the result is

$$\langle S\tau m | E_B | S\tau m\rangle$$

$$= \left[\frac{(\zeta+m)!(Z+m)!}{\zeta!\,m!\,Z!\,m!}\right]^{1/2} \frac{(n'-n-inn'K)^Z (n-n'-inn'K)^\zeta}{(n+n'+inn'K)^{\frac{1}{2}T+\frac{1}{2}\tau}}$$

$$\times (4nn')^{\frac{1}{2}m+\frac{1}{2}} \;{}_2F_1(-\zeta, -Z; m+1; y)$$

Multiplying these expressions gives $\langle \sigma\tau m | E | S\tau m\rangle$; as a check, $K = 0$ gives (6.20) with $\beta = n'/n$.

The remaining part of the calculation proceeds as in the preceding section. The matrix elements of $N + W_4$ are known from (E.1) and (6.19), giving factors in which m and $|m|$ are interchangeable:

$$4n\langle n'f'g' | e^{ikz} | nfg\rangle = 4n'\langle \sigma\tau m | E | S\tau m\rangle$$

$$-(\sigma+m-1)^{1/2}(\sigma-m-1)^{1/2}\langle \sigma-2 \ \tau \ m | E | S\tau m\rangle$$

$$-(\sigma+m+1)^{1/2}(\sigma-m+1)^{1/2}\langle \sigma+2 \ \tau \ m | E | S\tau m\rangle$$

$$+(\tau+m-1)^{1/2}(\tau-m-1)^{1/2}\langle \sigma \ \tau-2 \ m | E | S\tau m\rangle$$

$$+(\tau+m+1)^{1/2}(\tau-m+1)^{1/2}\langle \sigma \ \tau+2 \ m | E | S\tau m\rangle$$

After substituting the matrix elements of E, the number of different hypergeometric functions can be reduced by using the relation between contiguous functions as in Section 6.8. Define $\Phi_m(\lambda, v) = {}_2F_1(-\lambda, -v; |m| + 1; y)$ where $y = -4nn'/\{(n-n')^2 + K^2 n^2 n'^2\}$, so that Φ_m reduces to Ψ_m when $k = 0$. Using the quantum numbers n_1 and n_2, so that $n_1 + n_2 - n_1' - n_2' = n - n'$, the result can be written ($K = hk/p$, m means $|m|$):

$$\langle n'f'g' | e^{ikz} | nfg \rangle$$

$$= \frac{1}{2} K \left[\frac{(n_1' + m)!(n_2' + m)!(n_1 + m)!(n_2 + m)!}{n_1'! \, n_2'! \, n_1! \, n_2!} \right]^{1/2}$$

$$\times \left[\frac{4nn'}{(n + n')^2 + K^2 n^2 n'^2} \right]^{m+2}$$

$$\times \frac{(-)^{n_1' + n_2}(n - n' - iKnn')^{n_1 + n_2' - 1}(n - n' + iKnn')^{n_1' + n_2 - 1}}{m!^2 (n + n' - iKnn')^{n_1 + n_1'}(n + n' + iKnn')^{n_2 + n_2'}}$$

$$\times [\{2Kn^2 n'^2 + 2inn'(n_2 - n_1) + i(n^2 + n'^2 + K^2 n^2 n'^2)(n_1' - n_2')\}$$

$$\times \Phi_m(n_1, n_1') \Phi_m(n_2, n_2') + i\{(n + n')^2 + K^2 n^2 n'^2\}$$

$$\times \{n_2' \Phi_m(n_1, n_1') \Phi_m(n_2, n_2' - 1) - n_1' \Phi_m(n_1, n_1' - 1) \Phi_m(n_2, n_2')\}]$$

$$(6.29)$$

As a check the result $\langle nfg | e^{-ikz} | n'f'g' \rangle = \overline{\langle n'f'g' | e^{ikz} | nfg \rangle}$ may be verified by using the relation (O65)

$$\lambda \Phi_m(v, \lambda - 1) = v \Phi_m(\lambda, v - 1) + (\lambda - v) \Phi_m(\lambda, v)$$

Although the Φ_m are polynomials in y, the expression (6.29) has no negative powers of $\{(n - n')^2 + K^2 n^2 n'^2\}$, the denominator of y, because the degree of the $(n - n' \pm iKnn')$ terms is always greater than that of the Φ_m.

The matrix elements of z can now also be obtained by taking the coefficient of K in (6.29). The same method will give a formula for the matrix elements of any power of z.

6.10. SYMMETRIES OF MATRIX ELEMENTS

From Section 3.6, if an operator A anticommutes with the parity Π, then $\langle n'g'f' | A | ngf \rangle = (-)^{n - n' - 1} \langle n'f'g' | A | nfg \rangle$, as already noted for $A = p_x$. According to (5.16), interchanging f and g implies interchanging n_1 and n_2, so (6.23) clearly displays the result. Similarly $\Pi e^{ikz} = e^{-ikz}\Pi$ leads to

$$\langle n'g'f' | e^{-ikz} | ngf \rangle = (-)^{n - n'} \langle n'f'g' | e^{ikz} | nfg \rangle,$$

which is obvious from (6.29). For matrix elements between the states $u_{n_1 n_2 m}$ the phase difference $(-)^{n - n'}$ is replaced by $(-)^{m - m'}$ because (3.10) and (5.21) give $\Pi u_{n_1 n_2 m} = (-)^m u_{n_2 n_1 m}$. This is also clear from (5.3), and if the matrix elements are written down as integrals the symmetry follows by interchanging ξ and η.

In such integrals the substitution $\phi \to 2\pi - \phi$ relates matrix elements differing in the sign of m. This symmetry may be derived algebraically by the method at the end of Section 2.3 using $R = e^{-i\pi L_1}$. Putting

$$R = \exp(-i\pi F_x)\exp(-i\pi G_x)$$

shows that $R|nfg\rangle = i^{4F}|n \ -f \ -g\rangle$. From $Rz = -zR$, etc.,

$$\langle n' \ -f' \ -g'|p_x|n \ -f \ -g\rangle = (-)^{n-n'}\langle n'f'g'|p_x|nfg\rangle$$

$$\langle n' \ -f' \ -g'|p_z|n \ -f \ -g\rangle = (-)^{n-n'-1}\langle n'f'g'|p_z|nfg\rangle$$

$$\langle n' \ -f' \ -g'|e^{-ikz}|n \ -f \ -g\rangle = (-)^{n-n'}\langle n'f'g'|e^{ikz}|nfg\rangle$$

The first result extends (6.22) to states with $m < 0$, and the last two are consistent with (6.23) and (6.29), remembering that $|nfg\rangle \to |n \ -f \ -g\rangle$ means interchanging n_1 and n_2 as well as changing the sign of m.

These results are due to symmetries under reflection and rotation, but to get relations peculiar to the Coulomb problem the hidden symmetry must be used. For example, the invariance under geometrical inversion noted at the end of Chapter 4 allows the evaluation of the substitution $p' = p_0^2/p$ in radial integrals of momentum space wave functions. Hence

$$\langle nlm|f(p)|nlm\rangle = p_0^6\langle nlm|p^{-6} \ f(p_0^2/p)|nlm\rangle \tag{6.30}$$

for any f giving convergent integrals. In particular, taking $f(p) = 1$ and $f(p) = p^2$ leads to $\langle nlm|p^r|nlm\rangle = p_0^r (r = -6$ and $-8)$.

6.11. A TIME-DEPENDENT NONINVARIANCE ALGEBRA

An energy eigenfunction $R_{nm}e^{im\phi}$ of the two-dimensional oscillator is made time-dependent by multiplying by $e^{-in\omega t}$; then shift operators must also change this phase factor. For example, the $o_{2,1}$ subalgebra $\mathscr{M}$ of Section 6.6 should be redefined so that $T_{\pm} \leftrightarrow e^{\mp 2i\omega t}(U_3 \pm iV_3)$. If the domain of the algebra consists of solutions of the time-dependent Schrödinger equation, the Hamiltonian can be replaced by $i\hbar\partial/\partial t$, so that S can be replaced by $(i/2\omega)(\partial/\partial t)$. Since $U_3 \pm iV_3 = S - \frac{1}{2}\alpha\xi \pm \frac{1}{2} \pm \xi\partial/\partial\xi$, this substitution can also be made in the shift operators.

This procedure suggests defining a time-dependent $o_{2,1}$ noninvariance algebra by

$$T_3 = \frac{i}{2\omega}\frac{\partial}{\partial t},$$

$$T_1 = -i\sin 2\omega t\left(\frac{1}{2} + \xi\frac{\partial}{\partial\xi}\right) - \cos 2\omega t\left(\frac{1}{2}\alpha\xi - T_3\right)$$

$$T_2 = -i\cos 2\omega t\left(\frac{1}{2} + \xi\frac{\partial}{\partial\xi}\right) + \sin 2\omega t\left(\frac{1}{2}\alpha\xi - T_3\right)$$

$$\tag{6.31}$$

These operators are Hermitian if $\xi^{-1} - \partial/\partial\xi$ is the complex conjugate of $\partial/\partial\xi$, which requires the inner product to be defined by integration with respect to $\frac{1}{2}d\xi/\xi$ instead of $\frac{1}{2}d\xi$. The integral then diverges when $m = 0$, so that the domain of the operators (6.31) will be restricted to the odd parity states, with m odd and n even. To allow the application given in the next paragraph, the functions $f(\xi, \phi, t)$ of this domain should be the set satisfying the following conditions: (1) for constant t, f satisfies the usual conditions for a wave function, (2) $ih\,\partial f/\partial t = Hf$, and (3) f has period 2π in ϕ and period π/ω in t. All odd parity solutions of the time-dependent Schrödinger equation are included; in particular, the action of (6.31) on time-dependent energy eigenfunctions is still obtained by replacing $ih\partial/\partial t$ by the Hamiltonian, giving the same results as in Section 6.6. The representation of $o_{2,1}$ is therefore a direct sum of irreducible representations D_k^+ with $k = 1, 2, 3, \dots$ $(k = \frac{1}{2}|m| + \frac{1}{2}, a = \frac{1}{2}n - \frac{1}{2}|m| - \frac{1}{2})$.

The operators (6.31) give

$$[T_1, \xi^{-1}\sin 2\omega t] = i\xi^{-1}, \quad [T_1, \xi^{-1}\cos 2\omega t] = 0$$
$$[T_2, \xi^{-1}] = i\xi^{-1}\cos 2\omega t, \quad [T_2, \xi^{-1}\sin 2\omega t] = 0$$
$$[T_3, \xi^{-1}\cos 2\omega t] = -i\xi^{-1}\sin 2\omega t, \quad [T_3, \xi^{-1}] = 0$$

Comparison with (2.10) (with a sign change in the first equation) shows that $\xi^{-1}\cos 2\omega t$, $-\xi^{-1}\sin 2\omega t$, and ξ^{-1} are the Cartesian components of an $o_{2,1}$ vector with respect to (6.31). From the Wigner-Eckart theorem, their matrix elements have the same dependence on $k + a = \frac{1}{2}n$ (the eigenvalues of T_3) as the matrix elements of T_1, T_2, and T_3. Because of the modified inner product, these matrix elements of ξ^{-1} are physical expectation values of ξ^{-2}. Hence $\langle n|\xi^{-2}|n\rangle$ is just n multiplied by a factor dependent on $|m|$ only.

This theory was first obtained (A70) for Coulomb matrix elements in the $|nlm\rangle$ basis. As this work appeared after this book was completed, only a summary of it will be given. A representation of $o_{2,1}$ is defined by operators analogous to (6.31), with τ replacing $-2\omega t$, and the domain consisting of Coulomb wave functions multiplied by periodic functions of τ. (However, $-\tau/2\omega$ is not time in the Coulomb application.) The representation is Hermitian when the inner product is modified. For any integer k, operators $r^k e^{iq\tau}$ are components of tensor operators with respect to this algebra, and the Wigner-Eckart theorem shows that the expectation values of r^k depend on n through $o_{2,1}$ Clebsch-Gordan coefficients. Calculations reveal that the relevant coefficients depend on n in the same way that the familiar o_3 coefficients depend on m, that is, as in (2.9).

References

(A57)   S. P. Alliluev, *Zh. Eksp. Teor. Fiz.* **33**, 200 (1957); English transl.: *Sov. Phys.* *JETP* **6**, 156 (1958).

(A69)   R. H. Albert, *Proc. Camb. Phil Soc.* **65**, 107 (1969).

(A70)   L. Armstrong, *J. de Phys.* **31**, C4 (1970); *Phys. Rev.* **A3**, 1546 (1971); *J. Math. Phys.* **12**, 953 (1971).

(B36)   V. Bargmann, *Z. Phys.* **99**, 576 (1936). Bargmann's $D_{\rho\sigma}$ would be $D_{\sigma\rho}$ using the definition (1.7), and his A_k would be $-A_k/P$ using the definition (3.3). See also (Sc65).

(B47)   V. Bargmann, *Ann. Math.* **48**, 568 (1947).

(B57)   H. A. Bethe and E. E. Salpeter, *Quantum Mechanics of One and Two Electron Systems* (Springer-Verlag, Berlin, 1957); also in *Atoms I*, Vol. 35, *Handbuch der Physik* (Springer-Verlag, Berlin, 1957) S. Flugge (Ed.).

(B61)   L. C. Biedenharn, *J. Math. Phys.* **2**, 433 (1961). Biedenharn's $D_{\alpha\beta}$, **A**, **L**, **M**, and **N** become $iD_{\alpha\beta}$, **M**, **L**, **K**, and **N** in my notation.

(B65)   A. O. Barut, *Lectures in Theoretical Physics* (Gordon and Breach, New York, 1967), Vol. IXA.

Barut's notation: (taking $\lambda = i$) L_{13} L_{23} L_{12} $L^{\pm}$ ξ_1 ξ_2

My notation: T_1 T_2 T_3 $\sqrt{\tfrac{1}{2}}T_{\pm}$ z w

The article is an expansion of *Proc. Roy. Soc.* **A287**, 532 (1965), (with C. Fronsdal).

(BB65)   L. C. Biedenharn and P. Brussaard, *Coulomb Excitation* (Clarendon, Oxford, 1965).

(BI66)   M. Bander and C. Itzykson, *Rev. Mod. Phys.* **38**, 330 (1966).

(BW66)   S. Bell and P. A. Warsop, *J. Mol. Spectrosc.* **20**, 425 (1966).

Their notation: α r^2 l u v' v'' β $\psi_{u,\ \pm l}$

My notation: α ξ $|m|$ $a+|m|$ $n-1$ $N-1$ $1/\beta^2$ $(-)^m R_{nm} e^{\pm im\phi}$

See also *J. Phys.* **B3**, 745 (1970), in which the Laguerre polynomial has a different normalization.

(B67)   H. Bacry, *Lectures in Theoretical Physics* (Gordon and Breach, New York, 1967), Vol. IXA, p. 111.

(BK67)   A. O. Barut and H. Kleinert, *Phys. Rev.* **156**, 1541 (1967).

Their notation: **J**   **K** j_3 k_3 N_1^+ N_1^- N_2^+ N_2^- A_+^+ A_+^- A_-^+ A_-^-

My notation: **F**   **G** f g $-Q_0^\dagger$ $-Q_0$ $P_0^\dagger$ P_0 $Q_+^\dagger$ $-Q_-$ $-Q_-^\dagger$ Q_+

Their notation: L_{23}, L_{31}, L_{12} L_{i4} L_{i5} L_{i6} L_{56}

Mine

(Appendix D): L_1 L_2 L_3 M_i V_i W_i N

The "2" in their equation (A5) appears to be a misprint.

(BK67a)   A. O. Barut and H. Kleinert, *Phys. Rev.* **157**, 1180 (1967). Their equations (1.2) are equivalent to (E.1) in Appendix E through the correspondence $(\phi \to 0)$: $A_- \leftrightarrow -a_2$, $A_+ \leftrightarrow b_1$, $B_- \leftrightarrow b_2$, $B_+ \leftrightarrow a_1$, $* \leftrightarrow \dagger$.

(BK67b)   A. O. Barut and H. Kleinert, *Phys. Rev.* **160**, 1149 (1967).

(C16)   J. L. Coolidge, *A Treatise on the Circle and the Sphere* (Clarendon Press, Oxford, 1916), p. 21.

(C69)   A. Cisneros and H. V. McIntosh, *J. Math. Phys.* **10**, 277 (1969).

(D63)   Yu. N. Demkov, *Zh. Eksp. Teor. Fiz.* **44**, 2007 (1963); English transl.: *Sov. Phys. JETP* **17**, 1349 (1963).

(E34)   L. P. Eisenhart, *Ann. Math.* **35**, 284 (1934).

(E60)   A. R. Edmonds, *Angular Momentum in Quantum Mechanics*, 2nd ed. (Princeton University Press, 1960), p. 21.

(E62)   E. Eriksen, *Phys. Norv.* **1**, 121 (1962).

(F35)   V. Fock, *Z. Phys.* **98**, 145 (1935). See also (Sc65).

(F66)   G. Flamand, *J. Math. Phys.* **7**, 1924 (1966). Flamand's **A**, **K**, **F**, **G**, and Ω are $-\mathbf{A}/hP$, $-\mathbf{A}_n$, **G**, **F**, and $-[\mathbf{B}, H]/4E_n$ in my notation.

(F69)   P. L. Ferreira, *Rev. Mex. Fis.* **18**, 233 (1969).

(G29)   W. Gordon, *Ann. Physik* **2**, 1031 (1929).

(G59)   S. Goshen and H. J. Lipkin, *Ann. Phys. (N.Y.)* **6**, 301 (1959).

(G64)   M. L. Goldberger and K. M. Watson, *Collision Theory* (Wiley, New York, 1964), p. 10.

(G68)   S. Goshen and H. J. Lipkin, *Spectroscopic and Group-Theoretical Methods in Physics* (North-Holland, Amsterdam, 1968), F. Bloch et al. (Eds.), p. 255.

(H32)   E. Hylleraas, *Z. Phys.* **74**, 216 (1932).

(H33)   L. Hulthen, *Z. Phys.* **86**, 21 (1933). See also (Sc65).

(H65)   U. W. Hochstrasser, *Orthogonal Polynomials* in *Handbook of Mathematical Functions* (Dover, New York, 1965), M. Abramowitz and I. A. Stegun (Eds.).

(H66)   R. C. Hwa and J. Nuyts, *Phys. Rev.* **145**, 1188 (1966).

(H67)   J. W. B. Hughes, *Proc. Phys. Soc.* **91**, 810 (1967), who uses X_i and Y_i for F_i and G_i.

(H68)   J. W. B. Hughes, private communication.

(I51)   L. Infeld and T. E. Hull, *Rev. Mod. Phys.* **23**, 31 (1951).

(J40)   J. M. Jauch and E. L. Hill, *Phys. Rev.* **57**, 644 (1940).

(J62)   N. Jacobson, *Lie Algebras* (Wiley, New York, 1962).

(K33)   O. Klein, footnote on p. 22 in (H33).

(K62)   B. Kursunoglu, *Modern Quantum Theory* (Freeman, San Francisco, 1962).

(K66)   J. J. Klein, *Am. J. Phys.* **34**, 1039 (1966).

(K68)   H. Kleinert, *Lectures in Theoretical Physics* (Gordon and Breach, New York, 1968), Vol. XB. The notation used is that of (BK67) and (BK67a); also Kleinert's operators $N_1{}^1$, $N_1{}^2$, $N_1{}^3$, $N_2{}^1$, $N_2{}^2$, $N_2{}^3$ are $-A_1$, $-A_2$, A_3, $-B_1$, $-B_2$, B_3 in Section 6.9.

(L59)   P. O. Lowdin, *Advan. Chem. Phys.* **2**, 219 (1959).

(L60)   J. D. Louck and W. H. Shaffer, *J. Mol. Spectrosc.* **4**, 285 (1960).

(L66)   H. J. Lipkin, *Lie Groups for Pedestrians*, 2nd ed. (North-Holland, Amsterdam, 1966), Section 5.6.

(L66D)   According to Lipkin (L66, Appendix D) confusion cannot be avoided!

(M43–54)   W. Magnus and F. Oberhettinger, *Functions of Mathematical Physics* (Chelsea, New York, 1954), translated from the 1943 German edition.

(M58)   A. Messiah, *Mecanique Quantique* (Dunod, Paris, 1958); English transl. (North Holland, Amsterdam, 1961).

(MM65)   I. A. Malkin and V. I. Man'ko, *Zh. Eksp. Teor. Fiz., Pis'ma Redaktsiyu* **2**, 230 (1965); English transl.: *JETP Lett.* **2**, 146 (1966). See also *Yad. Fiz.* **3**, 372 (1966); English transl.: *Sov. J. Nucl. Phys.* **3**, 267 (1966).

(M66)   R. Musto, *Phys. Rev.* **148**, 1274 (1966), in which $\mathbf{A}$ is $-2\mathbf{A}_n$.

(O65)   F. Oberhettinger, *Hypergeometric Functions*, in *Handbook of Mathematical Functions* (Dover, New York, 1965), M. Abramowitz and I. A. Stegun (Eds.), especially (15.2.13) and (15.2.14), p. 558.

(P26–67) W. Pauli, *Z. Phys.* **36**, 336 (1926); English transl.: *Sources of Quantum Mechanics* (North-Holland, Amsterdam, 1967), B. L. van der Waerden (Ed.).

(P29)   B. Podolsky and L. Pauling, *Phys. Rev.* **34**, 109 (1929). See also the footnote on p. 378 of (S65a).

(P60)   D. Park, *Z. Phys.* **159**, 155 (1960). My interest in this subject was originally aroused by this paper, which closed a gap of twenty-four years in the literature.

(P66)   R. H. Pratt and T. F. Jordan, *Phys. Rev.* **148**, 1276 (1966), in which M_{J4}, $\mathbf{J}_1$, and $\mathbf{J}_2$ are $-2A_{nj}$, $\mathbf{F}$, and $\mathbf{G}$.

(R55)   M. E. Rose, *Multipole Fields* (Wiley, New York, 1955), p. 92.

(R57)   M. E. Rose, *Elementary Theory of Angular Momentum* (Wiley, New York, 1957).

(R65)   J. Roberts, M.Sc. thesis (University of Wales, 1965), unpublished.

(R67)   F. Ravndal and T. Toyoda, *Nucl. Phys.* **B3**, 312 (1967). Their operators $a_\pm$, $b_\pm$, $\mathbf{X}$, and $\mathbf{Y}$ correspond to $-iA_\pm$, $-iB_\pm$, $\mathbf{F}$, and $\mathbf{G}$.

(S44)   W. H. Shaffer, *Rev. Mod. Phys.* **16**, 253 (1944).

(S56)   A. P. Stone, *Proc. Camb. Phil. Soc.* **52**, 424 (1956). Stone's coordinates θ' and χ are $\tfrac{1}{2}\pi - \xi$ and η in (1.17) and Appendix C.

(S60)   J. L. Synge, *Classical Dynamics*, Vol. 3/1, *Handbuch der Physik* (Springer-Verlag Berlin, 1960), S. Flugge (Ed.), pp. 17–18. The χ and $\mathbf{U}$ are θ and $\mathbf{n}$ in Section 1.7.

(S65)   T. Shibuya and C. E. Wulfman, *Am. J. Phys.* **33**, 570 (1965).

(S65a)   T. Shibuya and C. E. Wulfman, *Proc. Roy. Soc.* **A286**, 376 (1965).

(Sc65)   D. E. Schwalm, Ph.D. thesis (University of Colorado, 1965, published by University Microfilms, Ann Arbor), p. 148. Also, Chapter 1 contains a unified account of (P26–67), (H33), (F35), and (B36).

(S68)   L. I. Schiff, *Quantum Mechanics*, 3rd ed. (McGraw-Hill, New York, 1968).

(S68a)   N. V. V. J. Swamy and C. V. Sheth, *Nuov. Cim.* **56A**, 29 (1968).

(S70)   N. V. V. J. Swamy, R. G. Kulkarni, and L. C. Biedenharn, *J. Math. Phys.* **11** 1165 (1970).

(T68)   J. D. Talman, *Special Functions—a Group Theoretic Approach* (Benjamin, New York, 1968), p. 185.

(T70)   C. B. Tarter, *J. Math. Phys.* **11**, 3192 (1970).

(W31–59) E. P. Wigner, *Gruppentheorie* (Braunschweig, 1931). English transl.: *Group Theory* (Academic Press, New York, 1959), p. 191.

(Z67)   B. Zaslow and M. E. Zandler, *Am. J. Phys.* **35**, 1118 (1967).

Appendix A. Evaluation of Commutation Relations

The basic commutators $[p_x, f(x, y, z)] = -ih\,\partial f/\partial x$, etc., and $[\mathbf{r}, f(\mathbf{p})] = ih\nabla f(\mathbf{p})$ lead to

$$[\mathbf{r}, p^2] = 2ih\mathbf{p}, \quad [\mathbf{p}\cdot\mathbf{r}, p^2] = 2ihp^2 \tag{A.1}$$

$$[\mathbf{p}, r^{-1}] = ih\mathbf{r}r^{-3}, \quad [\mathbf{p}, r^{-3}] = 3ih\mathbf{r}r^{-5} \tag{A.2}$$

$$\mathbf{r}\cdot\mathbf{p} - \mathbf{p}\cdot\mathbf{r} = 3ih, \quad [\mathbf{r}, (\mathbf{p}\cdot\mathbf{r})] = [\mathbf{r}, (\mathbf{r}\cdot\mathbf{p})] = ih\mathbf{r} \tag{A.3}$$

where each vector result stands for three equations, one for each component. As an example of the use of the vector results, $[p^2, r^{-1}]$ can be evaluated by writing

$$p^2 r^{-1} = \mathbf{p}\cdot\mathbf{p}r^{-1} = \mathbf{p}\cdot(r^{-1}\mathbf{p} + ih\mathbf{r}r^{-3}) \quad \text{using (A.2)}$$
$$= \mathbf{p}r^{-1}\cdot\mathbf{p} + ih(\mathbf{p}\cdot\mathbf{r})r^{-3}$$

and

$$\mathbf{p}r^{-1}\cdot\mathbf{p} = (r^{-1}\mathbf{p} + ih\mathbf{r}r^{-3})\cdot\mathbf{p} \quad \text{using (A.2)}$$

giving

$$[p^2, r^{-1}] = ihr^{-3}(\mathbf{r}\cdot\mathbf{p}) + ih(\mathbf{p}\cdot\mathbf{r})r^{-3} \tag{A.4}$$

Similarly (A.2) gives

$$(\mathbf{p}\cdot\mathbf{r})\mathbf{p}r^{-1} = (\mathbf{p}\cdot\mathbf{r})(r^{-1}\mathbf{p} + ih\mathbf{r}r^{-3})$$
$$= (\mathbf{p}r^{-1}\cdot\mathbf{r})\mathbf{p} + ih(\mathbf{p}\cdot\mathbf{r})\mathbf{r}r^{-3}$$

and again using (A.2)

$$\mathbf{p}r^{-1}\cdot\mathbf{r} = (r^{-1}\mathbf{p} + ih\mathbf{r}r^{-3})\cdot\mathbf{r} = r^{-1}\mathbf{p}\cdot\mathbf{r} + ihr^{-1}$$

so that

$$[(\mathbf{p}\cdot\mathbf{r})\mathbf{p}, r^{-1}] = ihr^{-1}\mathbf{p} + ih(\mathbf{p}\cdot\mathbf{r})\mathbf{r}r^{-3} \tag{A.5}$$

Now (3.3) can be written $h\mathbf{A} = p^2\mathbf{r} - (\mathbf{p}\cdot\mathbf{r})\mathbf{p} - hPr r^{-1}$, and (A.4) and (A.5) give

$$[\mathbf{A}, r^{-1}] = ir^{-3}(\mathbf{r}\cdot\mathbf{p})\mathbf{r} - ir^{-1}\mathbf{p} \tag{A.6}$$

From (A.1), $[p^2\mathbf{r} - (\mathbf{p} \cdot \mathbf{r})\mathbf{p}, p^2] = 0$, hence

$$[\mathbf{A}, p^2] = -P[\mathbf{r}r^{-1}, p^2] = -P[r^{-1}, p^2]\mathbf{r} - Pr^{-1}[\mathbf{r}, p^2]$$
$$= ihP(-2r^{-1}\mathbf{p} + r^{-3}\{\mathbf{r} \cdot \mathbf{p} + \mathbf{p} \cdot \mathbf{r} + 3ih\}\mathbf{r}) \qquad \text{using (A.4),}$$
$$\text{(A.1), and (A.2)}$$
$$= 2hP[\mathbf{A}, r^{-1}] \qquad \text{using (A.3) and (A.6)}$$

Hence $[\mathbf{A}, 2\mu H] = [\mathbf{A}, p^2] - [\mathbf{A}, 2hPr^{-1}] = 0$; that is, $\mathbf{A}$ is a constant of the motion.

The commutators between components of $\mathbf{A}$ and $\mathbf{L}$ are direct consequences of those between components of $\mathbf{r}$ or $\mathbf{p}$ and $\mathbf{L}$, because $\mathbf{L}$ commutes with p^2, $\mathbf{p} \cdot \mathbf{r}$, and r^{-1}, and the relation $\mathbf{A} \cdot \mathbf{L} = 0$ follows in the same way.

Using (A.1) and also $[\mathbf{p} \cdot \mathbf{r}, \mathbf{p}] = ih\mathbf{p}$, $h\mathbf{A} \times h\mathbf{A}$ becomes

$$\{p^2\mathbf{r} - (\mathbf{p} \cdot \mathbf{r})\mathbf{p} - hPr^{-1}\mathbf{r}\} \times \{\mathbf{r}p^2 - \mathbf{p}(\mathbf{p} \cdot \mathbf{r}) - 3ih\mathbf{p} - hPr\mathbf{r}^{-1}\}$$
$$= \{-p^2(\mathbf{p} \cdot \mathbf{r}) - 3ihp^2 + (\mathbf{p} \cdot \mathbf{r})p^2 - hP(\mathbf{p} \cdot \mathbf{r})r^{-1} + hPr^{-1}(\mathbf{p} \cdot \mathbf{r})$$
$$+ 3ih^2Pr^{-1}\}h\mathbf{L}$$

Using (A.1) and $[\mathbf{p} \cdot \mathbf{r}, r^{-1}] = ihr^{-1}$ gives $\mathbf{A} \times \mathbf{A} = -2i\mu H\mathbf{L}$.
Since $\mathbf{p} \times \mathbf{L} = -\mathbf{L} \times \mathbf{p} + 2i\mathbf{p}$,

$$A^2 = \mathbf{A} \cdot \mathbf{A} = (\mathbf{p} \times \mathbf{L} - i\mathbf{p} - Pr^{-1}\mathbf{r}) \cdot (\mathbf{p} \times \mathbf{L} - i\mathbf{p} - Pr\mathbf{r}^{-1})$$

and

$$\mathbf{p} \cdot (\mathbf{p} \times \mathbf{L}) = (\mathbf{p} \times \mathbf{p}) \cdot \mathbf{L} = 0, \mathbf{p} \times \mathbf{L} \cdot \mathbf{p} = (-\mathbf{L} \times \mathbf{p} + 2i\mathbf{p}) \cdot \mathbf{p} = 2ip^2,$$
$$\mathbf{p} \times \mathbf{L} \cdot \mathbf{r} = (-\mathbf{L} \times \mathbf{p} + 2i\mathbf{p}) \cdot \mathbf{r} = hL^2 + 2i\mathbf{p} \cdot \mathbf{r}, \mathbf{r} \cdot (\mathbf{p} \times \mathbf{L}) = \mathbf{r} \times \mathbf{p} \cdot \mathbf{L} = hL^2$$

Writing $(\mathbf{p} \times \mathbf{L})^2$ out in components, and using (2.10), shows that $(\mathbf{p} \times \mathbf{L})^2 = p^2L^2 - (\mathbf{p} \cdot \mathbf{L})^2 = p^2L^2$. Finally,

$$A^2 = p^2L^2 + 2p^2 - hPr^{-1}L^2 - 2iP(\mathbf{p} \cdot \mathbf{r})r^{-1} - p^2$$
$$+ iP(\mathbf{p} \cdot \mathbf{r})r^{-1} - hPr^{-1}L^2 + iPr^{-1}(\mathbf{r} \cdot \mathbf{p}) + P^2$$

Using (A.3) and $[\mathbf{r} \cdot \mathbf{p}, r^{-1}] = ihr^{-1}$ gives

$$A^2 = p^2L^2 + p^2 - 2hPr^{-1}L^2 - 2hPr^{-1} + P^2 = P^2 + 2\mu H(L^2 + 1)$$

Similar procedures give the results

$$\mathbf{A} \cdot \mathbf{r} - \mathbf{r} \cdot \mathbf{A} = i(\mathbf{r} \cdot \mathbf{p} + \mathbf{p} \cdot \mathbf{r}), \mathbf{A} \cdot \mathbf{p} - \mathbf{p} \cdot \mathbf{A} = i(p^2 + 2\mu H) \qquad \text{(A.7)}$$
$$\mathbf{A} \times \mathbf{r} = (\mathbf{p} \cdot \mathbf{r})\mathbf{L}, \mathbf{r} \times \mathbf{A} = -(\mathbf{r} \cdot \mathbf{p})\mathbf{L} \qquad \text{(A.8)}$$
$$\mathbf{A} \times \mathbf{p} = -\mathbf{p} \times \mathbf{A} = (p^2 + 2\mu H)\mathbf{L} \qquad \text{(A.9)}$$

Appendix **B.** Semisimple Lie Algebras

This appendix defines ideals, simple algebras, semisimple algebras, and structure constants. The result given in (1.6) of Chapter 1 holds for simple and semisimple algebras.

A subalgebra $\mathscr{S}$ of a Lie algebra $\mathscr{L}$ is an ideal if $[A, B]$ is in $\mathscr{S}$ whenever B is in $\mathscr{S}$ for all A in $\mathscr{L}$. For example, the three-dimensional algebra e_2 defined by the Lie products $[X, Y] = 0$, $[L, X] = Y$, $[Y, L] = X$ contains an ideal spanned by X and Y. Another example is provided by (1.11) which shows that o_3 subalgebras spanned by $\mathbf{K}$ and $\mathbf{N}$ are ideals of o_4. According to Section 1.15, an ideal is a subalgebra which is invariant under $\mathscr{L}$.

Every algebra has zero and itself as trivial ideals. If there are no others, the algebra is simple. A Lie algebra is Abelian if all its Lie products are zero. An algebra is semisimple if it is not Abelian and has no Abelian ideals other than zero. Thus vector algebra is simple, o_4 is semisimple, and e_2 is neither simple nor semisimple, since its ideal is Abelian.

Suppose that E_i $(i = 1, 2, \ldots, n)$ are a basis for an algebra, and

$$[E_i, E_j] = \sum_k c_{ij}^k E_k$$

Then the c_{ij}^k are called a set of structure constants for the algebra. Define $g_{pq} = \sum_{ik} c_{pi}^k c_{qk}^i$. Then the algebra is semisimple (or simple) if the nth order determinant with elements g_{pq} is not zero. One example of a Casimir operator is always given by

$$\sum_{i,j=1}^{n} g_{ij} E_i E_j$$

In the representation of o_3 given by angular momentum components J_x, J_y, and J_z, with the Lie product defined as (commutator)$/i$, the structure constants are $c_{xy}^z = -c_{yx}^z = 1$, $c_{yy}^x = 0$, etc. So $g_{pq} = -2\delta_{pq}$, $|g_{pq}| = -8 \neq 0$, and $-2J^2$ commutes with any component.

Appendix C. Explicit Expressions for Bases of o_4

Spherical polar coordinates for a four-dimensional space are defined by

$$x_1 = R \sin \alpha \sin \theta \cos \phi, \qquad x_2 = R \sin \alpha \sin \theta \sin \phi$$
$$x_3 = R \sin \alpha \cos \theta, \qquad x_4 = R \cos \alpha \tag{C.1}$$

Then (L60, K62)

$$D_{12} = -\frac{\partial}{\partial \phi}, \qquad D_{23} = \cos \phi \cot \theta \frac{\partial}{\partial \phi} + \sin \phi \frac{\partial}{\partial \theta}$$

$$D_{31} = \sin \phi \cot \theta \frac{\partial}{\partial \phi} - \cos \phi \frac{\partial}{\partial \theta}$$

$$D_{14} = -\frac{\sin \phi}{\sin \theta} \cot \alpha \frac{\partial}{\partial \phi} + \cos \theta \cos \phi \cot \alpha \frac{\partial}{\partial \theta}$$

$$+ \cos \phi \sin \theta \frac{\partial}{\partial \alpha} \tag{C.2}$$

$$D_{24} = \frac{\cos \phi}{\sin \theta} \cot \alpha \frac{\partial}{\partial \phi} + \cos \theta \sin \phi \cot \alpha \frac{\partial}{\partial \theta}$$

$$+ \sin \phi \sin \theta \frac{\partial}{\partial \alpha}$$

$$D_{34} = -\sin \theta \cot \alpha \frac{\partial}{\partial \theta} + \cos \theta \frac{\partial}{\partial \alpha}$$

and the Casimir operator is

$$L^2 + M^2 = -\frac{\partial^2}{\partial \alpha^2} - 2 \cot \alpha \frac{\partial}{\partial \alpha}$$

$$- \operatorname{cosec}^2 \alpha \left(\frac{\partial^2}{\partial \theta^2} + \cot \theta \frac{\partial}{\partial \theta} + \operatorname{cosec}^2 \theta \frac{\partial^2}{\partial \phi^2} \right) \tag{C.3}$$

Alternatively, using the coordinates defined by

$$x_1 = R \cos \xi \cos \phi, \qquad x_2 = R \cos \xi \sin \phi$$
$$x_3 = R \sin \xi \cos \eta, \qquad x_4 = R \sin \xi \sin \eta \qquad \text{(C.4)}$$

the basis of $c * o_4$ used in (1.11) has the form

$$K_z = -\frac{1}{2} i \left(\frac{\partial}{\partial \phi} + \frac{\partial}{\partial \eta} \right), \qquad N_z = \frac{1}{2} i \left(\frac{\partial}{\partial \eta} - \frac{\partial}{\partial \phi} \right)$$

$$K_\pm = \frac{1}{2} e^{\pm i\phi \pm i\eta} \left(\mp \frac{\partial}{\partial \xi} - i \cot \xi \, \frac{\partial}{\partial \eta} + i \tan \xi \, \frac{\partial}{\partial \phi} \right) \qquad \text{(C.5)}$$

$$N_\pm = \frac{1}{2} e^{\pm i\phi \mp i\eta} \left(\mp \frac{\partial}{\partial \xi} + i \cot \xi \, \frac{\partial}{\partial \eta} + i \tan \xi \, \frac{\partial}{\partial \phi} \right)$$

and the Casimir operator is

$$4K^2 = -\frac{\partial^2}{\partial \xi^2} - \sec^2 \xi \, \frac{\partial^2}{\partial \phi^2} - \text{cosec}^2 \, \xi \, \frac{\partial^2}{\partial \eta^2} + (\tan \xi - \cot \xi) \frac{\partial}{\partial \xi}$$

$$\text{(C.6)}$$

Appendix D. Hermitian Bases for $o_{3,2}$ and $o_{4,2}$

The tables list (commutator)/i. (This could be taken as the definition of the Lie product and real algebras of Hermitian operators used.) For example, the entry in the third row and fifth column of Table 1 means $L_3 U_2 - U_2 L_3 = -iU_1$. The relation to the skew-Hermitian basis (1.33) is $L_1 = iD_{23}$, etc., $U_\alpha = iE_{\alpha 4}$, $V_\alpha = iE_{\alpha 5}$, $S = iD_{45}$.

Table 1. Commutators defining $o_{3,2}$ and showing $o_{3,1}$ and $o_{2,2}$ subalgebras

	L_1	L_2	L_3	U_1	U_2	U_3	V_1	V_2	V_3	S
L_1		L_3	$-L_2$	0	U_3	$-U_2$	0	V_3	$-V_2$	0
L_2			L_1	$-U_3$	0	U_1	$-V_3$	0	V_1	0
L_3				U_2	$-U_1$	0	V_2	$-V_1$	0	0
U_1					$-L_3$	L_2	$-S$	0	0	$-V_1$
U_2						$-L_1$	0	$-S$	0	$-V_2$
U_3							0	0	$-S$	$-V_3$
V_1								$-L_3$	L_2	U_1
V_2									$-L_1$	U_2
V_3										U_3

Omitting S and the V_α gives $o_{3,1}$; omitting L_2, L_3, U_1, and V_1 gives an $o_{2,2}$ algebra; while S, U_3, and V_3 give an $o_{2,1}$ algebra.

In Table 2 opposite, L_α and M_α give the subalgebra o_4. Omitting W_α and N gives the de Sitter subalgebra $o_{4,1}$. Omitting L_2, L_3, M_1, V_1, and W_1 gives an $o_{3,2}$ subalgebra, the correspondence with Table 1 being as follows:

Table 1: L_1 $\quad L_2$ $\quad L_3$ $\quad U_1$ $\quad U_2$ $\quad U_3$ $\quad V_1$ $\quad V_2$ $\quad V_3$ $\quad S$

Table 2: M_3 $\quad -M_2$ $\quad L_1$ $\quad V_2$ $\quad V_3$ $\quad V_4$ $\quad W_2$ $\quad W_3$ $\quad W_4$ $\quad -N$

Also omitting (cf. Table 1) M_2, L_1, V_2, and W_2 gives the $o_{2,2}$ subalgebra used in Section 6.9, which generates transformations of x_3, x_4, x_5, and x_6, where x_i are the coordinates in the six-dimensional hyperspace transformed by the $SO_{4,2}$ pseudorotations.

Table 2. Commutators defining $o_{4,2}$

	L_1	L_2	L_3	M_1	M_2	M_3	V_1	V_2	V_3	V_4	W_1	W_2	W_3	W_4	N
L_1		L_3	$-L_2$	0	M_3	$-M_2$	0	V_3	$-V_2$	0	0	W_3	$-W_2$	0	0
L_2			L_1	$-M_3$	0	M_1	$-V_3$	0	V_1	0	$-W_3$	0	W_1	0	0
L_3				M_2	$-M_1$	0	V_2	$-V_1$	0	0	W_2	$-W_1$	0	0	0
M_1					L_3	$-L_2$	V_4	0	0	$-V_1$	W_4	0	0	$-W_1$	0
M_2						L_1	0	V_4	0	$-V_2$	0	W_4	0	$-W_2$	0
M_3							0	0	V_4	$-V_3$	0	0	W_4	$-W_3$	0
V_1								$-L_3$	L_2	$-M_1$	N	0	0	0	W_1
V_2									$-L_1$	$-M_2$	0	N	0	0	W_2
V_3										$-M_3$	0	0	N	0	W_3
V_4											0	0	0	N	W_4
W_1												$-L_3$	L_2	$-M_1$	$-V_1$
W_2													$-L_1$	$-M_2$	$-V_2$
W_3														$-M_3$	$-V_3$
W_4															$-V_4$

For example, $W_1 W_2 - W_2 W_1 = -iL_3$.

Appendix E. The Coulomb Noninvariance Algebra

BASIS OF THE OSCILLATOR REPRESENTATION OF $o_{4,2}$ (E.1)

$$L_1 \leftrightarrow -\tfrac{1}{2}(A_- e^{-i\phi}B_+^* + A_-^* e^{i\phi}B_+ - A_+^* e^{-i\phi}B_- - A_+ e^{i\phi}B_-^*)$$

$$L_2 \leftrightarrow \tfrac{1}{2}i(A_- e^{-i\phi}B_+^* - A_-^* e^{i\phi}B_+ - A_+^* e^{-i\phi}B_- + A_+ e^{i\phi}B_-^*)$$

$$L_3 \leftrightarrow \tfrac{1}{2}(A_+^* A_+ - A_-^* A_- + B_+^* B_+ - B_-^* B_-)$$

$$M_1 \leftrightarrow \tfrac{1}{2}(A_- e^{-i\phi}B_+^* + A_-^* e^{i\phi}B_+ + A_+^* e^{-i\phi}B_- + A_+ e^{i\phi}B_-^*)$$

$$M_2 \leftrightarrow \tfrac{1}{2}i(-A_- e^{-i\phi}B_+^* + A_-^* e^{i\phi}B_+ - A_+^* e^{-i\phi}B_- + A_+ e^{i\phi}B_-^*)$$

$$M_3 \leftrightarrow \tfrac{1}{2}(A_+^* A_+ + A_-^* A_- - B_+^* B_+ - B_-^* B_-)$$

$$V_1 \leftrightarrow \tfrac{1}{2}(A_+ e^{i\phi}B_+ + A_+^* e^{-i\phi}B_+^* + A_- e^{-i\phi}B_- + A_-^* e^{i\phi}B_-^*)$$

$$V_2 \leftrightarrow \tfrac{1}{2}i(A_+ e^{i\phi}B_+ - A_+^* e^{-i\phi}B_+^* - A_- e^{-i\phi}B_- + A_-^* e^{i\phi}B_-^*)$$

$$V_3 \leftrightarrow \tfrac{1}{2}(A_+^* A_-^* + A_+ A_- - B_+^* B_-^* - B_+ B_-)$$

$$V_4 \leftrightarrow \tfrac{1}{2}i(A_+ A_- - A_+^* A_-^* + B_+ B_- - B_+^* B_-^*)$$

$$W_1 \leftrightarrow \tfrac{1}{2}i(A_+^* e^{-i\phi}B_+^* - A_+ e^{i\phi}B_+ + A_-^* e^{i\phi}B_-^* - A_- e^{-i\phi}B_-)$$

$$W_2 \leftrightarrow \tfrac{1}{2}(A_+^* e^{-i\phi}B_+^* + A_+ e^{i\phi}B_+ - A_-^* e^{i\phi}B_-^* - A_- e^{-i\phi}B_-)$$

$$W_3 \leftrightarrow \tfrac{1}{2}i(A_+^* A_-^* - A_+ A_- - B_+^* B_-^* + B_+ B_-)$$

$$W_4 \leftrightarrow \tfrac{1}{2}(A_+^* A_-^* + A_+ A_- + B_+^* B_-^* + B_+ B_-)$$

$$N \leftrightarrow \tfrac{1}{2}(A_+^* A_+ + A_-^* A_- + B_+^* B_+ + B_-^* B_- + 2)$$

THE BASIS AS DIFFERENTIAL OPERATORS (E.2)

$$L_1 \leftrightarrow i(\xi\eta)^{-1/2}[\xi\eta(D_\eta - D_\xi)\sin\phi + \tfrac{1}{2}(\xi - \eta)\cos\phi\, D_\phi]$$

$$L_2 \leftrightarrow i(\xi\eta)^{-1/2}[(D_\xi - D_\eta)\cos\phi + \tfrac{1}{2}(\xi - \eta)\sin\phi\, D_\phi]$$

116

$$L_3 \leftrightarrow -iD_\phi$$

$$M_1 \leftrightarrow (4\alpha^2 \xi\eta)^{-1/2}[-\xi\eta(4D_\xi D_\eta - \alpha^2)\cos\phi$$
$$+ \cos\phi\, D_\phi^2 + 2(\xi D_\xi + \eta D_\eta)\sin\phi\, D_\phi]$$

$$M_2 \leftrightarrow (4\alpha^2 \xi\eta)^{-1/2}[-\xi\eta(4D_\xi D_\eta - \alpha^2)\sin\phi$$
$$+ \sin\phi\, D_\phi^2 - 2(\xi D_\xi + \eta D_\eta)\cos\phi\, D_\phi]$$

$$M_3 \leftrightarrow (4\alpha)^{-1}[\alpha^2(\xi - \eta) - 4(D_\xi - D_\eta + \xi D_\xi^2 - \eta D_\eta^2) - (\xi^{-1} - \eta^{-1})D_\phi^2]$$

$$V_1 \leftrightarrow (4\alpha^2 \xi\eta)^{-1/2}[\xi\eta(4D_\xi D_\eta + \alpha^2)\cos\phi$$
$$- \cos\phi\, D_\phi^2 - 2(\xi D_\xi + \eta D_\eta)\sin\phi\, D_\phi]$$

$$V_2 \leftrightarrow (4\alpha^2 \xi\eta)^{-1/2}[\xi\eta(4D_\xi D_\eta + \alpha^2)\sin\phi - \sin\phi\, D_\phi^2$$
$$+ 2(\xi D_\xi + \eta D_\eta)\cos\phi\, D_\phi]$$

$$V_3 \leftrightarrow (4\alpha)^{-1}[\alpha^2(\xi - \eta) + 4(D_\xi - D_\eta + \xi D_\xi^2 - \eta D_\eta^2) + (\xi^{-1} - \eta^{-1})D_\phi^2]$$

$$V_4 \leftrightarrow i(\xi D_\xi + \eta D_\eta + 1)$$

$$W_1 \leftrightarrow i(\xi\eta)^{-1/2}[-\xi\eta(D_\xi + D_\eta)\cos\phi + \tfrac{1}{2}(\xi + \eta)\sin\phi\, D_\phi]$$

$$W_2 \leftrightarrow -i(\xi\eta)^{-1/2}[\xi\eta(D_\xi + D_\eta)\sin\phi + \tfrac{1}{2}(\xi + \eta)\cos\phi\, D_\phi]$$

$$W_3 \leftrightarrow i(\eta D_\eta - \xi D_\xi)$$

$$W_4 \leftrightarrow (4\alpha)^{-1}[\alpha^2(\xi + \eta) + 4(D_\xi + D_\eta + \xi D_\xi^2 + \eta D_\eta^2) + (\xi^{-1} + \eta^{-1})D_\phi^2]$$

$$N \leftrightarrow (4\alpha)^{-1}[\alpha^2(\xi + \eta) - 4(D_\xi + D_\eta + \xi D_\xi^2 + \eta D_\eta^2) - (\xi^{-1} + \eta^{-1})D_\phi^2]$$

in which $D_\xi = \partial/\partial\xi$, $D_\eta = \partial/\partial\eta$, and $D_\phi = \partial/\partial\phi$.

For the hydrogen atom, Cartesian and parabolic coordinates are related by

$$x = (\xi\eta)^{1/2}\cos\phi, \qquad y = (\xi\eta)^{1/2}\sin\phi, \qquad z = \tfrac{1}{2}\xi - \tfrac{1}{2}\eta \qquad \text{(E.3)}$$

$$D_x = \frac{2x}{\xi + \eta}(D_\xi + D_\eta) - \frac{y}{\xi\eta}D_\phi, \qquad D_y = \frac{2y}{\xi + \eta}(D_\xi + D_\eta) + \frac{x}{\xi\eta}D_\phi,$$
$$\text{(E.4)}$$

$$(\xi + \eta)D_z = 2\xi D_\xi - 2\eta D_\eta$$

Index